THE CALEDONIAN CANAL

D1324543

To
The Men and Women
Who Work the Canal

THE CALEDONIAN CANAL

Fourth Edition

A.D. Cameron

BIRLINN

This edition first published in 2005 by
Birlinn Limited
West Newington House
10 Newington Road
Edinburgh
EH9 1QS

www.birlinn.co.uk

2

Copyright © A.D. Cameron 1972, 1983, 1994, 2005

The right of A.D. Cameron to be identified as the author of
this work has been asserted by him in accordance with
the Copyright, Designs and Patents Act 1988

All rights reserved. No part of this publication may be
reproduced, stored, or transmitted in any form, or by
any means, electronic, mechanical or photocopying,
recording or otherwise, without the express
written permission of the publisher.

ISBN: 978 1 84158 403 4

British Library Cataloguing-in-Publication Data
A catalogue record for this book is available
from the British Library

Typesetting and prepress origination by Brinnoven, Livingston
Printed and bound by Gutenberg Press, Malta

CONTENTS

LIST OF ILLUSTRATIONS

INTRODUCTION AND ACKNOWLEDGEMENTS TO THE FIRST EDITION

As would be expected of a public enterprise, documents about the construction and operation of the Caledonian Canal are abundant. The most useful collections are the Ministry of Transport Records in the National Archives of Scotland in Edinburgh, and the Papers on the Caledonian Canal etc. in the House of Lords Record Office. Written and printed sources which have been valuable are listed in the bibliography. The Canal workers themselves have been another mine of information. To them, Canal boats like *Scot* and *Wee Jean* are real personalities and the Canal itself a living tradition. It has been a privilege and a delight to be welcomed within their fold because of an interest shared with them, to have questions willingly answered and to listen to tales well told.

I wish to express my gratitude for the assistance given to me in Edinburgh by the staffs of the National Archives of Scotland, the National Library of Scotland and the Scottish Room of the Public Library; in London in the House of Lords Record Office, the British Museum and the Institution of Civil Engineers; and in the north in Inverness Public Library, the Forestry Commission and Tourist Offices in Fort William and Inverness. I wish to acknowledge my particular indebtedness to Mr Brian Davenport, Engineer Scotland, of the British Waterways Board, who first suggested that I should write this book and who advised on engineering aspects of the Canal. I also wish to thank the British Waterways Board for making records and photographs available to me in the Canal Office; *The Highland News*, the Scottish Tourist Board, the Highlands and Islands Development Board and Mr Jim Hogan of Caley Cruisers for help in finding interesting photographs; and Mr Douglas Stuart who drew the maps.

INTRODUCTION AND ACKNOWLEDGEMENTS TO THE FOURTH EDITION

This book was published in October 1972 to coincide with the Canal's 150th Anniversary. For that reason the description of the Canal and how it normally operated in 1972 has been left to stand as a record of that time. Elsewhere in the book, however, the text has been revised after examining recent works and the evidence which only came to light in 1992, in the form of maps, plans, wage books and masses of typed material. Centralised and being listed in the new Canal Office on Muirtown Basin, it was unknown to me as author in 1971–2 and apparently to the Canal management at the time, who had given me open access to all the records in the old Canal Office at Clachnaharry and every encouragement to make full use of them. These Canal records are now in the Highland Council Archive in Inverness.

About half of the records discovered in 1992 had lain out-of-sight for decades in a building which had formerly been the Canal manager's stables and something like a third had been piling up on successive district inspectors over the years in the middle and western districts of the Canal. The number of very old plans discovered is not large but some do add to or illuminate the construction story and I am very grateful to Mr Hugh Ross, the Canal manager at that time, for giving permission to display some of them here and to Mr John Ross for helping me to sift through them. I also wish to thank Ailsa Andrews and Nigel Rix of the Canal staff for their help and encouragement towards the publication of this new edition.

Special thanks are due to my son David Cameron, Milestone Systems, Avonbridge, for preparing the text of the revision and Mrs Robina McGregor for her immaculate typing. Mr Alec Howie and Beatrice Clark of British Waterways helped by arranging for modern photographs. Mr Ian Gowans of Robert H. Cuthbertson and Partners kindly explained his firm's recent works on the Canal and my friend Mr Frank Spaven helped me in updating Appendix II. A chapter has been added to deal with recent events on the Canal and the changes in traffic along it up to the present day. For permission to reproduce illustrations thanks go to the following: *Aerofilms Ltd.*, page 63; British Waterways, pages xiv, 25, 31, 35, 38, 56, 57, 65, 84, 86, 92, 97, 104, 112, 130, 131, 138, 141, 193, 208; Caley Cruisers, 186; David Cameron, 180; Beatrice Clark, 183; Paul Colenso, 182; Robert H. Cuthbertson and Partners, 177; John Dewar Studios, 16, 162; Glasgow Art Gallery, Kelvingrove, 168; Gordon Harvey, 118, 158 (right); *The Highland News*, 8, 12, 13, 43, 72, 106, 148, 151, 152; The Highlands and Islands Development Board, 9 (bottom), 81, 82, 119, 145 (top), 189; heirs of the late Col. P.F. Hone and Edinburgh Public Libraries, 30; The Institution of Civil Engineers, 28; Lady Lever Art Gallery, Port Sunlight, 19; The National Galleries of Scotland, 76; The National Library of Scotland, 33, 54, 70, 126, 166; The National Library of Wales, 5; Edmund Nuttall Ltd, 174; the late Brian Peach, 3, 49, 55, 95, 109, 149, 153, 155, 163 (left); Keeper of the Archives of Scotland, 123; *The Scotsman*, 22; Scottish Canals, 191; Scottish Ethnological Archive, National Museums of Scotland, 108; University of St Andrews Library, 133; Valentine's of Dundee and Edinburgh Public Libraries, 7. The other illustrations are the author's.

The Caledonian Canal

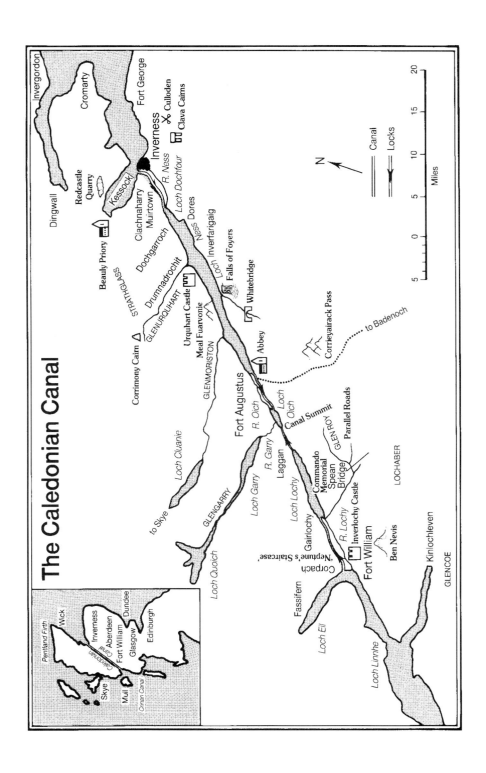

NOTICE.

Caledonian Canal.

SHIP-OWNERS, MASTERS, and BROKERS, are hereby informed, that the Rate of Tonnage Duty levied on Vessels passing through the CALEDONIAN CANAL, was reduced ONE HALF on the 1st of January, 1828, viz. :—From One Halfpenny per Ton per Mile, to One Farthing per Ton per Mile ; or from 2s. 7d. to 1s. 3½d. per Register Ton, for passing wholly through from Sea to Sea, a distance of Sixty-two Miles. Thus, any Vessel or Steamboat, laden or unladen, in Ballast, or with any Cargo whatever, say of 100 Tons Burthen, passing from Sea to Sea in either direction, now pays Dues to the amount of £6. 9s. 2d. No duty is charged upon any description of Goods.

The Depth of Water in the Navigation throughout, is FIFTEEN FEET ; and the Locks are 170 feet long, by 40 feet in width. The Voyage through the Canal with a fair wind, is generally performed by sailing vessels in less than two days ; and is seldom, under any circumstances, known to exceed a week.—No expense is necessary for Pilotage, and but a trifling sum for Tracking by horses in case of adverse winds. Every facility is afforded by the Officers and Lock-keepers of the Establishment in assisting Vessels through the Locks, &c.

The CALEDONIAN CANAL is found to afford important facilities for Vessels trading from the Eastern Coasts of the Kingdom to every part of Ireland, as well as to Glasgow, Liverpool, and all the Ports on the Western Coast of Scotland and England, as far as the Bristol Channel. It cuts off, in the former communication between these points, a distance of at least THREE HUNDRED MILES along the most difficult and dangerous Coast of the Island, in circumnavigating which, not only weeks, but months, it is well known, have frequently been occupied ; and while a manifest saving in point of time is thereby obtained, the risk is at the same time so far diminished as to do away with the necessity of Insurance both on Vessel and Cargo. The same advantages are held out to Ships voyaging from opposite quarters of the Kingdom to the Ports of the Baltic on one hand, and to America and the West Indies on the other ;—in all which cases, the low Rate of Tonnage Duty on the Canal, and the security it affords to the Property as well as lives of individuals, renders it an object of especial importance to Masters and Owners of Vessels trading in these directions, and to the Shipping interest of the Nation in general.

Caledonian Canal Office, 16th November, 1829.

The first hand-bill printed and distributed by George May on becoming Engineer in 1829.

1
THE OPENING OF THE CANAL

The Caledonian Canal was ceremonially opened when the first ship sailed through from Inverness to Fort William on 23–24 October 1822. On board were Mr Charles Grant, former MP for the County representing the Canal Commissioners, and an invited party of landed proprietors. The event was reported in the *Inverness Courier* in dignified prose, appropriate to the time and the occasion:

> The doubters, the grumblers, the prophets and the sneerers, were all put to silence, or to shame; for the 24th of October was at length to witness the Western joined to the Eastern sea. Amid the hearty cheers of the crowd of Spectators assembled to witness the embarkation, and a salute from all the guns that could be mustered, the Voyagers departed from the Muirtown Locks (Inverness) at 11 o'clock on Wednesday with fine weather and in high spirits. In their progress through this beautiful Navigation they were joined from time to time by the Proprietors on both sides of the lakes; and as the neighbouring hamlets poured forth their inhabitants, at every inlet and promontory, tributary groups from the glens and the braes were stationed to behold the welcome pageant, and add their lively cheers to the thunder of the guns and the music of the Inverness-shire militia band, which accompanied the expedition . . . [At Glen Urquhart,] where a number of Highlanders were gathered, the Voyagers were joined by Mr Grant of Redcastle, Mr Grant of Corriemony, the Rev. Mr Smith of Urquhart and several other gentlemen. The reverberation of the firing, repeated and prolonged by a thousand echoes from the surrounding hills, glens and rocks – the martial music – the shouts of the Highlanders – and the answering cheers of the party on board, produced an effect which will not soon be forgotten by those present . . .

Other stops and salutes consumed more time and it took all of seven hours to reach Fort Augustus to the noise of more guns, more music

Tall ships entering Laggan Locks from Loch Lochy in July 1991.

Urquhart Castle on the most commanding site overlooking Loch Ness.

and more congratulations. On Thursday, the party departed at six in the morning:

> After sailing about five and a half miles in the Canal and passing through seven locks, the steam yacht entered Loch Oich. On approaching the mansion of Glengarry, the band struck up 'My name it is Donald Macdonald' etc. and a salute was fired in honour of the Chief, which was returned from the old castle, the now tenantless residence of Glengarry's ancestors. The Ladies of the family stood in front of the modern mansion waving their handkerchiefs . . . The Voyagers were here joined by the *Comet (II)* steam-yacht . . . After passing through two locks, and a small portion of the Canal cut through the summit from which the land falls towards the East and West Sea, the yacht entered Loch Lochy . . . The groups of Highlanders (for all the huts of Lochaber must have been deserted), stationed upon picturesque and commanding points, added not a little to the interest and liveliness of the scene . . . The last portion of the Canal was now entered. It is

eight miles in length and contains 12 locks. At Banavie, near Corpach, eight of these grand locks, which are close upon each other, have been fancifully denominated 'Neptune's Staircase'. It was half past five when the vessel at last dipped her keel into the waters of the Western Ocean, amidst the loud acclamations of her passengers and a great concourse of spectators! The termination of the voyage was marked by a grand salute from the Fort, whilst the Inhabitants of Fort William demonstrated their joy by kindling a large bonfire. A plentiful supply of whisky, given by the gentlemen of Fort William, did not in the least dampen the ardour of the populace. 67 gentlemen, the guests of Mr Grant, sat down to a handsome and plentiful dinner . . .

In this manner the Caledonian Canal was opened 'from sea to sea'. The excitement of the spectators on shore was unrestrained because they had lived through the years of hard work that had gone into constructing it. Probably many of them had worked on it themselves, and they were delighted that it had that day proved itself serviceable. They had high hopes that it would be a safe route for ships for many years to come and that with ships would come trade and industry and a prosperous future.

What is significant about the opening party was that it consisted almost entirely of landed proprietors. They were still a power in the Highlands. Charles Grant must have had more than half the county's eighty-three voters aboard, since only landowners had the right to vote at that time. He was unique amongst them in having been personally involved in promoting the Canal, as one of the Commissioners since its inception. One of the company, Alastair MacDonell of Glengarry, of whom more later, had been a greater obstacle to the completion of the Canal than the stubborn rock from which Corpach basin was hewn. Their mood of mutual congratulation expanded after dinner when, after several loyal toasts which were not enumerated by the reporter, no less than thirty-nine toasts were proposed and drunk – a reminder that the tax on whisky was only 6/2d a gallon.

They drank to many things – the prosperity of the Canal, Parliament's generosity, the Chancellor of the Exchequer, the Canal Commissioners one by one, Chiefs and Clans, Inverness county and town – and to one another. It was not until the nineteenth toast that a word was said to praise the men who planned the Canal and the men who built it, when the Hon. William Fraser proposed 'Mr Telford, and the gentlemen who carried on the operative part of the Caledonian Canal with so much credit to themselves.'

At 12 o'clock the party broke up, the *Courier* reporter tells us, but some gentlemen, 'with genuine Highland spirit', carried on into the early hours of the morning. The next day they completed the voyage back to Inverness in thirteen hours. So much for Highland spirit!

The Canal had been opened in two days: it had taken nineteen years and cost nearly a million pounds to build.

Looking from Fort William towards Loch Eil. The Canal entrance was made at Corpach behind the island in the left centre. The Fort is to the right with the Great Glen beyond. From Pennant's *Tour in Scotland* (1769).

2
WHY BUILD A CANAL IN THE GREAT GLEN?

In a mountainous country with an indented coastline like Scotland, people had used the sea from the earliest times as the easiest means of communication. Most settlements were coastal and lines of penetration inland took advantage of firths and long sea lochs. The wealth of Bronze Age remains in the Great Glen, for example, shows that it attracted very early settlers, using river and loch to make their way inland and gain access to the east coast. Later, in the Middle Ages, most burghs in the Lowlands were created within the sound of the sea and traded with Europe and one another by sea. The Highlands, however, remained almost destitute of towns.

When the transport revolution came in England in the 1750s it started with the construction of canals to transport heavy goods like coal to the rising industrial towns. James Brindley's canal from inside the coal mine at Worsley to Manchester set the pattern. Direct delivery in bulk from an inland mine to an inland destination proved that inventiveness and hard work could cut costs and helped to make the Industrial Revolution possible. During the sixty years of the reign of George III the navigable stretches of English rivers were joined with one another by canals until Manchester was linked with London, and Gloucester, if need be, with Hull. Inland waterways became the arteries of trade, financed by private enterprise and cut by navigators, later called 'navvies'. Scotland, too, saw the construction of canals during this Canal Age.

The first project to be surveyed in Scotland was for a sea-to-sea canal across her narrow waist between the Forth and the Clyde. The second, the Monkland Canal, between Airdrie and Glasgow, was simply an artificial ditch dug with the aim of undercutting the coal prices charged by local mine-owners nearer to Glasgow. Work was going forward on both by 1770 and it was not surprising that thoughts were turning towards the prospect of cutting canals in many other parts of the country, even in the Highlands.

This is an example of how ideas for the improvement of a region at any time tend to follow current fashion. Projects which have proved successful earlier elsewhere are taken up and advocated with enthusiasm even where conditions are less favourable or, frankly, unfavourable. It was happening with estate improvement in the eighteenth century: the high returns received by improvers like Coke of Holkham in Norfolk were an incentive to others, but not something that was always achieved. It was happening with the foundation of factory villages, such as Stanley in Perthshire and Spinningdale in Sutherland, which had to compete with much larger units of production in Lancashire and Lanarkshire. It was to happen again much later with a proposal for a railway to Skye and the abortive railway between Spean Bridge and Fort Augustus which ran with few passengers along the south-western half of the Great Glen from 1903 to 1933 and was described by John Thomas in *The West Highland Railway* as 'the classic example of the railway which should not have been built'.

At the time the cry everywhere was for canals and the Great Glen running from south-west to north-east from Fort William to Inverness looked as if it had been shaped by nature to make the canal-builder's

Fishing boats in full sail by the back of Tomnahurich: the Brahan Seer's prophecy fulfilled.

job easy. Where else in mountainous country in Britain was there a stretch of sixty miles involving a rise above sea level of only about a hundred feet? Which other area contained three long narrow lochs in a straight line capable of forming two-thirds of the waterway without human effort? Where else was there such heavy rainfall to swell the mountain streams and keep the water level high? What loch could compare with Loch Ness in length (22 miles) or in depth (129 fathoms), deeper than any part of the North Sea between Scotland and Denmark? The location was attractive and the prospect of a great navigation there to join the Atlantic with the North Sea was to many observers an intriguing possibility.

Even in the seventeenth century the best-known Highland prophet, the Brahan Seer, predicted: 'strange as it may seem to you this day, the time will come, and it is not far off, when full-rigged ships will be seen sailing eastward and westward by the back of Tomnahurich [inland] at Inverness'.

Old field patterns in Glen Roy with pastures beyond, typical of former ways of farming in the Great Glen.

The Corrieyairack, the greatest of General Wade's military roads, from the Great Glen near Fort Augustus over towards Dalwhinnie, built in 1731 and being restored in 1992.

Cattle in Glen Nevis.

Serious proposals for canals in the Highlands, however, were not put forward simply to fulfil a prophecy or because a site looked promising. They were considered rather in the context of what might be called the Highland problem after the Jacobite debacle at Culloden in 1746. Following 'the harrying of the glens' and the people in them by Cumberland's Redcoats, the Government treated the Highlanders with studied repression. New laws forbade them to carry weapons, wear tartan or even play the bagpipes. In 1747 every chief lost his hereditary power to judge and punish members of his own clan and the estates of those chiefs who had supported the Jacobite cause were forfeited to the Crown and run by Commissioners. Government troops occupied the forts along the Great Glen and elsewhere, and the programme of building military roads, which General Wade started in 1721, was extended. These were all negative measures, aimed at preventing another rising or rebellion. But there was positive thinking too, shown best and earliest in the way the forfeited estates were administered.

Schemes of agricultural improvement and tree-planting were undertaken. Crafts for young people such as carpentry for boys and spinning for girls were encouraged and spinning wheels were provided. With the co-operation of the Society for Propagating Christian Knowledge, more schools were built, where children could learn to read and write English, although later the Society produced a Gaelic Bible. These were some of the ways in which the Commissioners tried to bring those areas of the Highlands for which they were responsible more into line with the rest of Britain. Theirs was the first attempt 'to do something about the Highlands', but their critics viewed their efforts as a campaign to anglicise the Highlanders and destroy a rich indigenous culture.

Broadly the Highlanders had differed from people further south in two main ways – their social organisation and their subsistence economy. The social structure had been based on the clan, subject to the absolute authority of the chief, held together by kinship and loyalty and designed to provide him with the greatest number of warriors. Between the chief and the tenants stood the tacksmen, who were usually his close relations and who held land from him on 'tack' or lease, at a low rent and lived off the rents and services paid to them by their tenants in return for their patches of land. The tacksmen might not be rich by southern standards but they were gentlemen, the chief's deputies in their areas and responsible for raising their tenants as fighting men. The Highland chiefs were the only men in Britain capable of raising an armed force spontaneously, as they did in the Rising of 1745. Admittedly clanship was decaying in parts of the Highlands before this time and clan feuds had become uncommon but even after the abolition of the Heritable Jurisdictions in 1747, old ways and old loyalties persisted. An ancient social structure does not disappear at the stroke of a lawyer's pen.

Clan society was supported by a primitive subsistence economy, part arable but mainly pastoral, devoted to keeping cattle as well as a few sheep and goats. The surplus cattle were sold or driven south along the drove roads to the cattle trysts of Crieff or Falkirk. The price they fetched paid the rent: there was usually little left over. Highland agricultural methods served to provide a fairly high population with their basic needs in relative poverty. Years of crop failure brought hunger and death, but the introduction of the potato crop proved a life-saver in the Highlands. Not only did it provide an alternative food when grain crops failed, but it yielded a greater amount of food than the same area under grain. It was probably a major factor in keeping more people alive in the Highlands and consequently in supporting the startling rise in population in the later years of the eighteenth century.

Recording a rise of 40,000 in the forty-six years to 1801 in the Highland and Island counties – Argyll, Inverness, Ross and Cromarty, Sutherland, Caithness, Orkney and Shetland – the figures do not reveal the full extent of the increase because they do not take into account the substantial losses through emigration. Between 1768 and 1800 at least 20,000 people left the Highlands for the colonies in America alone, not counting those who had gone south in search of work.

Before the '45, a clan chief's power and position depended on the number of fighting men he could muster. In the pacified Highlands after it, having a large number of tenants was becoming an embarrassment to him as he ceased to be a chief and became a landlord. Consequently, tacksmen no longer deserved their privileged position. Since cattle were increasing in value there was a strong case for a landlord specialising in cattle-rearing himself or encouraging some of his tenants to do so in return for higher rents. Some tacksmen accepted these terms; others,

Sheep in a landscape shaped by ice – the Parallel Roads of Glen Roy.

feeling redundant in this new economic situation, chose to emigrate in the 1770s and took their tenants with them. By the end of the century, the price of wool rose spectacularly, sheep proved more profitable to keep than cattle and vast areas of the Highlands were given over to flocks of Blackfaced and later Cheviot sheep. The rents the incoming Border and English farmers could afford to pay were so high that the Highland tenants could not compete. This transition from farming for survival to farming for profit broke the ancient bonds of kinship and established a landlord-tenant relationship often with new tenants who were able to pay a higher rent.

Some modern writers maintain that the Highlands were over-populated. Certainly, the Highlands contained more people in 1801 than they do today and a much larger proportion of the population of Scotland than today. But over-population is a relative term: it has to be considered in terms of the economy which sustains it. The Highlands were over-populated in terms of subsistence agriculture, where the

A gateway – formerly the main entrance to the original Fort William, 1690 –
re-erected on the spot where Alan Cameron of Erracht raised the Cameron
Highlanders over a century later.

shortage of arable land caused holdings to be split and people to be poorer. In terms of the new sheep-farming economy which was introduced, the area was over-populated because sheep farms needed fewer workers and destroyed the shielings which had previously been used for cattle.

There were two possible solutions: one to reduce population, either by encouraging emigration or military service, or second, to take steps to provide alternative occupations for the existing population. The Government was unwilling to subsidise emigration but William Pitt the Elder took steps to harness the military spirit of the Highlanders into service in the British Army. Although Duncan Forbes of Culloden had suggested this long before he did, Pitt boasted of his part in it:

> I was the first minister who looked for [merit] and found it in the mountains of the north. I called it forth and drew into your service a hardy and intrepid race of men. These men in the last war [the Seven Years War against France, 1756–63] were brought to combat on your side; they served with fidelity as they fought with valour and conquered for you in every part of the world.

From such beginnings the traditions of the Highland Regiments in the British Army developed but what was most remarkable was the willingness of members of clans with former Jacobite sympathies such as the Frasers and the Camerons to raise regiments to fight for the Crown. Those who saw the Highlands as a nursery for recruits for the army were unlikely to subscribe to new economic ideas aimed to provide the Highlanders with employment which would keep them at home.

Some enthusiasts thought the sea could supply the work and the food that the land could not. They formed the British Fisheries Society in 1786 with the aim of founding fishing villages on the west coast and the Islands as bases from which to fish for herring. Ullapool and Tobermory were two important settlements which they established. The connection between the Fisheries Society and proposals for the Caledonian Canal became important by 1800, first because Thomas Telford was their surveyor and engineer, having made a tour of the north-west and written a long report as early as 1790, second, because fishing villages needed better communications to get their catches to market and third, because the Society became a powerful means of communication with Government on Highland problems. For example, George Dempster of Skibo, MP, one of the Society's founders and one of Scotland's under-estimated patriots, was calling for Government expenditure on Highland roads in 1792:

> Do not forget our Highland roads. In the happy state of our finances, the judicious application of a few thousand on piercing the north and west Highlands with good roads would be a most patriotic application of public money and quickly repay the bountiful Treasury for its expenditure.

Crop failures in 1799 and 1800 caused hunger and destitution and a feeling of helplessness. A new wave of emigrants departed. Telford, on his investigation in 1802, reported, 'From the best information I have been able to procure, three thousand persons went away in the course of last year, and if I am rightly informed, three times that number are preparing to leave the country in the present year'. The British Fisheries Society too was receiving information on the extent and causes of emigration from agents like Dr William Porter at Lochbay

The Commando Memorial near Spean Bridge: Lochaber was a training ground for soldiers during the Second World War.

Looking down from Ben Nevis over Loch Eil, and the Caledonian Canal.

Fishing boats crowded in Clachnaharry Reach during the Kessock herring 'bonanza' in 1966, a reminder of the importance of herring to the Highland economy.

in Skye, but it also had reassuring evidence on the value of schemes to provide work:

> The fine winter has enabled us to carry on our roads with great spirit, which has furnished the people with constant employment and has kept all in this part of the country from the thought of emigrating. I shall have occasion to draw on you for £100 on account of the roads to complete the year.

Aware of the distress in the Highlands, the Government was alarmed at the loss of valuable manpower through emigration, especially in wartime. It had some information from the British Fisheries Society about the benefits of public works on a limited scale. Was it practicable to develop more harbours and cut a Canal through the Great Glen for use by fishing boats and ships of the Royal Navy? That was the problem on which Thomas Telford was asked to report in 1801.

3

THOMAS TELFORD: THE MAN WITH IDEAS

Through his connection as engineer to the British Fisheries Society, Thomas Telford was not unacquainted with the Highlands and their problems. He was not a Highlander but a Borderer, born in the parish of Westerkirk near Langholm in Dumfriesshire in 1757. His father was a shepherd who died when Thomas, an only child, was less than a year old. His education in the parish school had given him the rudiments of learning without ever stunting his natural curiosity and thirst for knowledge which were to be features of his life. He served his apprenticeship as a mason and in the early 1780s he was off to work first in the New Town of Edinburgh, then, having taken 'the high road to England', on Somerset House under Sir William Chambers in London, before moving to work on the Commissioner's House in Portsmouth Dockyard, where he laid aside his trowel and first rose to a position of responsibility. In 1787 he was in Shrewsbury planning to rebuild the Castle there for Sir William Pulteney. It is unlikely that this career of moving from job to job and place to place was planned. Partly it was the result of circumstances such as the reduction of spending on Somerset House which slowed up building there, but mainly it arose from the restless energy of a young man of talent, anxious to gain experience and get on in the world. Looked at in retrospect, it was as if, in an age when gentlemen went on the Grand Tour to complete their education, Thomas Telford the craftsman went on the little tour of the public works of Britain. He was extending his knowledge but his education was never 'completed': acquiring useful knowledge was a passion all his life.

So far, he was the builder who was growing into an architect. He had been making drawings and supervising alterations, particularly for

Opposite: Thomas Telford by Sir Henry Raeburn.

Sir William Pulteney, whose support proved very important at several points in Telford's career. Pulteney was in fact a Johnstone, the second son of the laird of Westerhall in Telford's native Eskdale, who changed his name to Pulteney when he married the heiress of the Earl of Bath. His marriage brought him great wealth and considerable influence in Shropshire, and he was Member of Parliament for Shrewsbury for nearly forty years. Pulteney's patronage helped Telford to become Surveyor of Public Works for Shropshire and in 1793 General Agent, Engineer and Overlooker to the Ellesmere Canal. This method of appointment was normal practice in the eighteenth century and there were other men of influence supporting other candidates. Patrons fortunately prided themselves on choosing and backing men of ability: and Telford was able. Without such help, however, it is unlikely that he would have risen so high or achieved so much. These posts extended his scope to town planning and bridge-building, including the third iron bridge in Britain at Buildwas over the Severn, and introduced him to the challenges of a new field, canal construction.

The Pontcysyllte Aqueduct carrying the Ellesmere Canal in an iron trough at great height over the Vale of Llangollen.

The consulting engineer on the Ellesmere Canal was William Jessop, who had been a pupil of John Smeaton, the surveyor and canal builder. Jessop had been the first secretary of the Society of Civil Engineers, later called the 'Smeatonians', which was founded in 1771 at the time when Telford was starting his apprenticeship as a mason. He had wide experience of waterways, such as the Grand Junction Canal and harbour works like the West India Docks and, although a modest and retiring man, his opinions were highly regarded by his fellows. Telford found him a tower of strength. Together they wrestled with their main problems, the carrying of the Ellesmere Canal by aqueducts across two broad and deep valleys, the Ceriog at Chirk and the Dee at Pontcysyllte. The Chirk Aqueduct, over 700 feet long and 70 feet high was massive in size compared with Brindley's pioneer project at Barton forty years earlier, and novel in having a canal bed of iron plates to reduce weight at such a height. This aqueduct was completed in 1801 and the other, over the Vale of Llangollen at Pontcysyllte, was well under way. It was even more challenging, for the valley was over two and a half thousand feet wide.

The length to be bridged was reduced to 1,000 feet by constructing a great mound of earth on the south bank. The canal was to be carried in iron troughs for lightness, supported by tall pillars rising to a height of over 120 feet above the River Dee. These engineering achievements were attracting attention far beyond the bounds of Shropshire while Telford's plan for an iron bridge to cross the Thames with a single span of 600 feet startled London. The bridge was not built, partly because of the technical difficulties and its sheer scale but also because of shortage of money. It was at this point in his career, with one engineering marvel just completed, another in the making and a third rejected, that he was called by the Government for an engineer's objective opinion on the problems of Highland communications.

This new task in his native country was a challenge to him and he was so keen to start that he was off to the Highlands before he received his detailed instructions. The extent of his travels and the way he drove himself on are revealed in a letter he wrote from Peterhead in October 1801 once his Highland tour was behind him:

> I have carried regular Surveys along the Rainy West through the tempestuous wilds of Lochaber, on each side of the habitation of the far-famed Johnny Groats, around the shores of Cromarty, Inverness and Fort George, and likewise the Coast of Murray [Moray]. The

apprehension of the weather changing for the worse has prompted me to incessant and hard labour so that I am now almost lame and blind; I have, however, I trust, now nearly accomplished all the main objects of my mission, and shall be able to make out a plan and surveys of one of the noblest projects that ever was laid before a Nation, the whole of which I am satisfied is practicable, at a given expense.

He need not have worried about 'the weather changing for the worse' for he was blessed with good weather and achieved far more than he had thought possible. So preoccupied was he with his Highland survey that he had not even time on his return to visit his native Langholm. Back in Shrewsbury in November, he had the problems of the Ellesmere Canal to catch up on, but every spare moment he had that winter were spent on his plans for the Highlands. 'Never when awake – and perhaps not always when asleep – have my Scotch Surveys been absent', he wrote to Andrew Little in Langholm in April 1802.

The Government's intentions were fourfold: to find promising fishing stations, to see if a canal from the east to the west coast was practicable, to find suitable harbours in the north-east for trade with the Baltic and use by the Royal Navy, and to establish 'a safe and convenient intercourse between the Mainland of Scotland and the Islands'.

In a preliminary reply to the Treasury, written at Peterhead in October 1801 (his 'Scotch Survey' does not survive), Telford paid little attention to the Government's last objective beyond recommending that Oban harbour be improved for that purpose. By substituting the general aim of improving communications between the different parts of England, Scotland and Ireland, he was able to knit the Government's objectives into a coherent whole. Central to his plan was the cutting of the Caledonian Canal, about which he had no doubts:

> I am convinced that a navigation may be formed from Inverness to Fort William of the description that is wanted. The line is very direct, and I have observed no serious obstacle in any part of it. The whole rise between the shores is trifling, and on the summit there is an inexhaustible supply of water. The Entrances from the Navigation into

Opposite: Fort George built after the Battle of Culloden to replace the Fort of the same name in Inverness, blown up by the Jacobites.

the Sea, are immediately in deep water, good anchoring ground and in places of perfect safety at Inverness and Fort William.

Oban and Cromarty could be developed into harbours offering protection and shelter at either end, as well as bases for naval escort vessels. By sailing through the Canal, ships would escape the dangers of navigating the Pentland Firth and the attentions of French privateers in wartime, while reaching their destinations more quickly. Ports in Ireland and the west of England would be brought into closer touch with ports on the east coast and the Baltic. Herring fishing would be promoted by building a harbour on the east coast at Wick and the Canal would allow fishing crews to sail quickly and safely from one coast to the other in search of the migrating herring shoals.

Telford's plan was not the first proposal for a canal through the Great Glen but it was unique in that it was commissioned directly by the Treasury. In earlier years, others had been busy promoting the idea. James Watt, famous for his improvements of the steam engine, had surveyed the route for the Commissioners of the Forfeited Estates and estimated that a canal 10 feet deep, with thirty-two locks joining the three Lochs could be provided for £164,000. No action was taken on Watt's report but pressure to turn the Great Glen into an artificial waterway was building up. In his *View of the Highlands*, published in 1784, John Knox was so convinced of its advantages that he expressed astonishment that it had not been completed years ago. Ministers in Inverness and other parishes recommended it at length in the *Statistical Account of Scotland* in the early 1790s, one even going to the length of declaring that 'nature has left little to be done'. John Rennie, the famous bridge builder and engineer, made another survey in 1793. The British Fisheries Society had been campaigning for it for years.

Telford knew James Watt and consulted him about his early survey before publishing his own Report. He was impressed by the line taken by Watt, although his own plan was for a canal on a grander scale.

After reporting in 1802 Telford was asked to return to the Highlands that summer to consider suggestions such as the possibility that a canal might be cut between the Caledonian Canal and Loch Duich through Glen Shiel, and to examine for the first time the wider causes of emigration.

He rejected the idea of offshoot waterways from the proposed Caledonian Canal as 'altogether unadvisable'. After Mark Gwynn, commander of the Loch Ness galley, had taken soundings for him in all three Lochs, Telford was convinced that two of the three were navigable and the smallest, Loch Oich, could be deepened. He was surer than ever that a canal was feasible and reckoned that it could be constructed in seven years at a total cost £350,000.

He made proposals too for roads and bridges at the entrances to the Highlands and north and west of the Great Glen and gained the goodwill of landowners by inviting the opinions of the Highland Society. He asked them to consider the value of roads and bridges and the Canal to the economic life of the Highlands and the reasonableness of expecting landowners to pay half the cost of roads through their estates. They had no doubts on either point and echoed his conviction that such public works would check the spirit of emigration by providing work and money, which would in turn stimulate local industry. Their co-operation was important, not only because Telford based his estimates of the costs of roads and bridges on the assumption of a fifty per cent contribution from landowners but also because his proposals were to be submitted to Parliament, which consisted mainly of landowners.

Telford handled the emigration question with persuasive skill. He appreciated the 'push' influence of the extension of large-scale sheep-farming, which forced the people to move out, and the 'pull' of persuasive letters sent by others who had emigrated before them, and

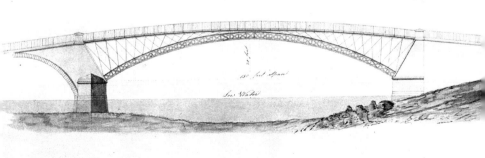

The original Bonar Bridge with its delicate iron arch, an example of Telford's other works in the Highlands, a drawing by Thomas Rhodes.

sometimes of deception 'by artful persons, who hesitate not to sacrifice these poor ignorant people to selfish ends'. He believed that the craze for sheep-farming was temporary, that prices of sheep and wool would fall when the Highlands were fully stocked, and that there would be a return to arable farming and cattle-rearing in the glens with sheep on the higher slopes only. This mixed farming, he thought, suited the Highlands best and was consistent with a high population but it would benefit from improved communications. If these roads and bridges had been constructed earlier, there would have been no emigration, or at least less. In view of the current flood of intending emigrants, therefore, he recommended:

> If there are any public works to be executed, which, when completed, will prove generally beneficial to the country, it is advisable these works should be undertaken at the present time. This would furnish work for the industrious and valuable part of the people in their own country, they would by this means be accustomed to labour, they would acquire some capital, and the foundations would be laid for future employments. If, as I have been credibly informed, the inhabitants are strongly attached to their native country, they would greedily embrace this opportunity of being able to remain in it, with the prospect of bettering their condition, because, before the works were completed, it must be evident to every one that the whole face of the country would be changed.

No government at that time would have considered paying relief to the unemployed destitute but the Government was prepared, in the special case of the Highlands, to lay out money to provide work for them on roads, bridges, harbours and the Caledonian Canal, in the hope that these would form an infrastructure which, by facilitating trade and the growth of industry, would provide more work for more people. Money given to help to complete the private enterprise Forth and Clyde Canal was a precedent for Government participation. In agreeing to pay the full cost of the Caledonian Canal and undertaking thereafter to run it, the Government launched what might be called the first nationalised enterprise.

No one estimated whether it would pay its way, or even asked the question. They accepted the assurances that it was feasible, that it would be useful and that work on it would help to stem the tide of emigration.

4

CHOOSING THE TEAM

On a plaque for the Pontcysyllte Aqueduct, the engineering marvel opened on 26 November 1805, are inscribed the names of the men who built it:

Thomas Telford was Engineer
Matthew Davidson was Superintendent of the Work
John Simpson and John Wilson executed the Masonry
William Hazledine executed the Iron Work
William Davies made the Earthen Embankment.

One name which ought to be there is missing, the name of William Jessop. Certainly his involvement in it was less in the last six years to 1805, when he himself was devoting over a hundred days a year to building the West India Docks in London, but Telford would have been the first to give Jessop full credit for his contribution in making the vital engineering decisions. Charles Hadfield challenged this view in his book *Thomas Telford's Temptation* in 1993, which is discussed further in Chapter 16.

Of these men, Jessop, Davidson, Simpson and Wilson, were called by Telford to help him in his grand design for the Highlands and came to work in the North. Hazledine, the ironmaster at Plas Kynaston near Pontcysyllte was to make all the ironwork for lock-gates and bridges at the west end of the Canal. Each was a specialist in his field, each was of proven ability – Pontcysyllte Aqueduct was a monument to their skill.

Civil engineering was then a relatively young profession and Telford had grown into it. By ingenuity and boldness he had solved new problems and found new answers to old problems. Those who worked closely with him – and he had a knack for recognising and attracting men of ability – rose to new responsibilities and challenges

and matured with him. An outstanding example was Matthew Davidson, a Dumfriesshire stone-mason like himself. He came to Shropshire at Telford's invitation as superintendent of bridge-building and became Inspector of Works on the Ellesmere Canal. Another, John Simpson, a Shrewsbury mason, gained experience in building bridges and according to Telford was a 'treasure of talents'. John Wilson, a mason from Dalston in Cumberland, and Thomas Davies and William Hughes who came north with him, proved to be formidable contractors.

Telford had the gift of commanding loyalty from his lieutenants. Hardly anyone ever left him and some, e.g. Davidson and Simpson, were with him all their lives. It is easy to argue that people worked for him because that was where the work was and where the money was but there were other civil engineers and other public works elsewhere – and the Highlands of Scotland had little to offer, compared with the milder climate of Wales, as Matthew Davidson was soon to discover.

Telford's plan for the Highlands was the biggest and most comprehensive programme of works for the development of any area ever undertaken in Britain. He was determined that it would be a professional operation, supervised by men with skill and a sense of responsibility. As far as the Canal was concerned he immediately secured the appointment of William Jessop as consulting engineer. Matthew Davidson became resident engineer at the north-east end at Inverness and John Telford, apparently no relation of Thomas, was brought from his job as toll-collector at Chester to take charge at Fort William. Each was to have £200 a year, a house and a horse. John Simpson was to be the contractor for all the mason work with John Wilson as his foreman in the west and John Cargill in the east. Canal-cutting and embankments on the other hand were to be let out in small lots to different people, or in larger lots to contractors with experience, the workmen having to provide their own picks, shovels and spades.

Similar care was taken to ensure a professional approach to the road building programme. Telford was fortunate that most of the proposed roads north and west of the Great Glen had been surveyed in the 1790s

Opposite: William Jessop, portrait copy by Edwin Williams which now hangs in the Institution of Civil Engineers; the original portrait is owned by the Jessop family..

Telford in a gig driven by John Rickman's son, while inspecting progress in the Highlands in 1819. From a watercolour reported lost at the time of the Telford bicentenary in 1957.

by George Brown, a skilled surveyor in Elgin. John Duncombe, who had been his assistant engineer on the Ellesmere Canal came north as Chief Inspector of Roads and ultimately had the help of six sub-inspectors. He needed them, not only because the area was vast but because he himself was to prove a failure in the Highlands and died, to Telford's disgust, 'in a jail at Inverness'. His successor, John Mitchell, however, did sterling work.

These were the key men on the spot. Their Parliamentary masters and paymasters in London were the Caledonian Canal Commissioners, and the Commissioners for Roads and Bridges. The Speaker of the House of Commons was chairman of both, the Chancellor of the Exchequer was a member of both, as were Charles Grant, the MP for Inverness-shire, and Sir William Pulteney, Telford's early patron. The Commissioners had to report progress and expenditure to Parliament each year.

Matthew Davidson, the Dumfriesshire mason who came north after completing the Pontcysyllte Aqueduct in North Wales to become resident engineer at Clachnaharry.

Acting as Secretary to both Commissions was John Rickman, a diligent public servant, who had suggested the first population census in 1801. James Hope, an Edinburgh lawyer, became their legal agent to deal with the purchase of land and claims for compensation. Their collaboration with Telford was to last much longer than they at first imagined.

Forming the link between these groups was Telford himself. He was for ever on the move: in the Highlands twice a year planning what was to be done and inspecting what had been done, calling on Hope in Edinburgh, writing reports to the Secretary from his office in Shrewsbury and attending meetings in London. Because of his frequent changes of station, he was asked to select dates convenient to himself for meetings with his controlling body, the Canal Commissioners.

Opposite: Designs for waggons built at Corpach and Clachnaharry for use on railways. Note the iron wheels and the shape of the rails. Fig. 1 is the elevation and Fig. 2 the side view of a back-tilting waggon, Fig. 7 a side-tilting waggon and Fig. 9 a waggon for carrying stones.

5

MAKING A START IN 1803

The scale of the Canal from sea to sea was determined by the size of ships that sailed the sea. It was to be a ship canal. The cutting was to be 15 feet deep with a bottom width of 50 feet and sloping sides. The excavated material would build up the banks to a level high enough to take 20 feet of water, giving the Canal a surface breadth of 110 feet. The proposed locks were to be 162 feet long and 38 feet wide, big enough to accommodate all the Baltic and West India vessels, but not the biggest frigates in the Royal Navy.

In August 1803, Telford came north to prepare for public works in an area which had never known them before, except for military forts and roads. He was entering the region of the north-west Highlands where the wheel was hardly known and the spoked wheel unknown. The sledge, the pack-horse and human shoulders were the commonest means of transporting what had to be moved on land. Farm implements were unsophisticated, wooden, home-made and short-lived. Small parts of them were of iron, like the sock of the foot-plough or the coulter of the horse-drawn plough on the flatter land in the east. He came among people with little arable land who had depended on their cattle and low rents for survival. They had almost no skills to offer him, only their hardiness and capacity to learn.

Telford set out the proposed line of the cutting and Jessop followed him, checking the line, digging trial pits to ascertain the nature of the soil, fixing the position of the locks and working out the first detailed estimate of cost. Rejecting Telford's idea of building locks with turf walls, which would cost less, and estimating for locks lined with stone, Jessop put the total figure at £474,531. This was excluding the cost of land, which he thought was of little value and which Telford valued at £15,000. The next summer they were asked to estimate the extra cost of making the Canal big enough to accommodate 32-gun and 44-gun frigates of the Royal Navy – a sensible question when Britain was at war against Napoleon's France, and of making smaller side locks for fishing vessels and coasters of the size which used the Crinan Canal. By increasing the size of the locks to 170 feet by 40 feet, and 180 feet by 40 feet where locks were built together, at an extra cost of £8,000, any of the forty-six 32-gun frigates in the Navy would be able to pass through without any need to widen the whole Canal. This was agreed and the idea of side locks was rejected as costly and unnecessary. £20,000 was allocated for preparatory work in 1803 and £50,000 a year after that.

The general plan for the Canal was clear in the minds of Jessop and Telford. The summit was at Laggan, between Loch Oich and Loch Lochy. Deep cutting would be necessary there, to bring it down to the level of the bed of Loch Oich. This Loch would require dredging to deepen it in places, but its water supply was ample. It was fed by the River Garry from the big mountain lochs, Loch Quoich and Loch Garry, in an area with the highest rainfall in the country. West of Laggan, Loch Lochy was to be dammed to raise its level by 12 feet,

Inside a lock – an idea of scale.

to avoid even deeper cutting between it and Loch Oich; a new outlet from the Loch was to be cut for the River Lochy to flow into the Spean, and the Canal outlet was to occupy the River's old bed. In the west, several aqueducts were necessary to carry the Canal over mountain streams feeding the Lochy. At Banavie, a flight of eight locks would be built together to save expense, lowering the level step by step right down to the cutting across Corpach Moss, to two locks above the Basin and the final sea-lock at Corpach.

Travelling eastwards from Loch Oich the Canal in places would occupy the bed of the River Oich on the way to Fort Augustus at the west end of Loch Ness. The siting of locks there was likely to prove difficult as the soil was a mixture of gravel and sand and water simply ran through it. It might be necessary to cut a new outlet for the River Oich through rock and use the river channel. Also, a canal 20 feet deep would require the foundation of the lowest lock to be built 24 feet below the summer level of Loch Ness.

At the east end of the Loch dredging would be necessary to deepen the channel of the River Ness where it enters Loch Dochfour and a weir would have to be constructed at the east end of the little Loch to maintain it at the same level as Loch Ness. Eastwards towards Inverness, Canal and river were to march side by side until the Canal curved left round Torvean Hill to Muirtown. There, four locks would lower it to the great basin which was to be constructed on the left of the channel, into which the River Ness overflowed in flood and the sea penetrated at high tide (today on the line of Abban Street, Carse Road, Muirtown Basin). The basin would terminate in a lock, leading into a lower basin behind the fishermen's cottages at Clachnaharry, to be formed between two long mounds thrusting out like a great crab's claw into deeper water to allow exit through the sea-lock to the Beauly Firth and the North Sea. A deep bed of soft clay there would provide a good foundation. The first choice of sea-entrance east of Kessock Ferry pier, nearer Inverness harbour, was abandoned because the ground proved to be a mixture of sand and gravel. Along the whole of the south-east side of the cuttings, but not along the Lochs, a towpath was to be constructed to allow ships to be towed by trackers with horses.

On similar works elsewhere in Britain, the civil engineer's responsibilities were, first, to determine the line of the Canal and the location of the locks, and then to arrange for supervision and payment as work progressed. In the Highlands, however, Jessop and Telford and their lieutenants had many other duties. They had to train the Highlanders to become efficient workers. They had to determine what materials were needed and in what quantity; and know or find out where they could be obtained. They became inventors and designers of equipment, such as tools and waggons, which were to be made on the site. They were quartermasters, laying out new settlements with houses and workshops, sometimes even providing food and drink for the men. As A.R.B. Haldane has shown in his book, *New Ways Through the Glens*, Telford himself acted as paymaster, all the money granted by Parliament passing through his own bank account. It was like a military campaign which was to outlast the nation's struggle against Napoleon.

Jessop and Telford decided to start on the basins at either end and to work inland so that parts of the Canal and the Lochs could be used later for transporting material for construction work inland. Clachnaharry in the east and Corpach in the west became the headquarters from which the Canal works were conducted. In his official reports and in the accounts of expenditure, Telford always referred to operations at the east end and the west end as happening at Clachnaharry and Corpach, the little settlements nearest to the sea entrances. The 'Clachers' of Clachnaharry would have been amused to know that their 'Fishermen's Houses at a Place called Clacknacarry' were mentioned in an Act of Parliament of 1804 and that the Members of Parliament, both Lords and Commons, who were members of the Canal Commission, met in London from time to time to consider how things were in Clachnaharry.

The villagers could see what was happening for themselves. On the Inverness side of the village, workshops for carpenters and blacksmiths and a storehouse for tools were being erected near the new sheds and huts where the workmen lived. Men were already working at the new rubble quarry in the hillside just west of the village and across the Firth at Redcastle others were opening up a freestone quarry to obtain better-quality stone for building the locks. At Kessock Samuel Deadman was building a 50 ton sloop, to be called the *Caledonia*, to carry the stone from Redcastle to the new stone pier with the crane for unloading at Clachnaharry. An Aberdeen sea-captain, Murdoch Downie, had been out there in the Firth taking soundings and marking on a chart to establish the best approach to the proposed entrance to the Canal.

The place was no longer just a fishing village. It was filling up with tradesmen – masons, carpenters and smiths. The Canal seemed to be a vast undertaking, judging from the amount of material and equipment that was arriving from the south and all the waggons and implements which were being constructed on the site. Cargoes of tools and timber were coming in regularly, tools off a London ship and load after load of fine Baltic pine in ships from Aberdeen. Local landowners and timber merchants were supplying wood as well. Mr Duff of Muirtown had been cutting fir and birch on his estate nearby and Alexander Fraser, an Inverness merchant, had been buying great quantities in Glenmoriston and on the banks of Loch Ness and was sailing loads of it, oak and birch, in a 15 ton vessel along Loch Ness. This was much quicker than floating logs and allowed him also to deliver goods such as oatmeal to Fort Augustus on the return journey.

Smiths were making picks and crowbars while carpenters were

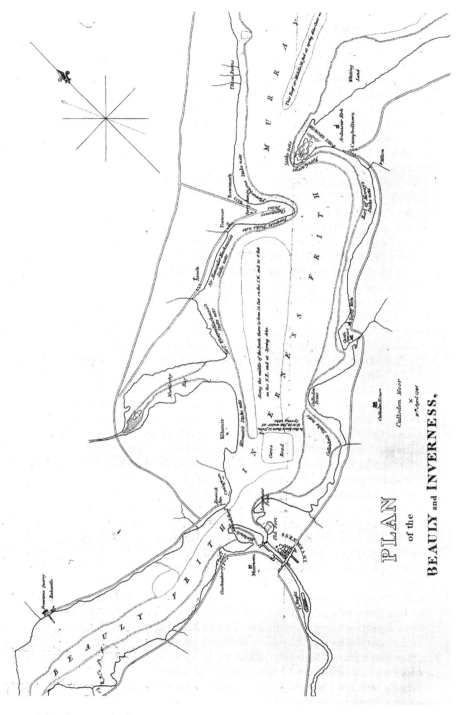

The plan Murdoch Downie made of the approach to the Canal at Inverness, showing his soundings, the sites of Kessock Ferry and Clachnaharry and the freestone quarry at Redcastle.

building wheel-barrows to Telford's specifications and a price of 15/2¼d. They were turning out all kinds of waggons with iron wheels. There were turn-up or tip-up waggons, some that tipped backwards, others that tipped sideways; there were gravel or drop-through waggons, and waggons with no sides for carrying stone. The estimated cost of materials and workmanship to make a gravel waggon was £11/17/11½d and a turn-up waggon £13/14/4d. Since each waggon was designed to carry one cubic yard of earth when loaded up almost to the top of the sides, the calculation of wages per cubic yard removed was simplified. Iron rails for making tramways were coming in by sea, mainly from Benjamin Outram & Co., a big ironworks at Butterley in Derbyshire, in which Jessop was a partner and probably provided the designs of the waggons on page 33. The new waggons were to be pulled by horses along these tramways or railways. There was word too that three big steam engines, one with the power of thirty-six horses, would soon be coming from the firm of Boulton & Watt in Birmingham. Clachnaharry had never seen anything like it. Neither had nearby Inverness.

The town was growing steadily in population with the coming of industries, using hemp and flax, to reinforce trade which had always been its main preoccupation. It was an important port, with seven vessels of over 400 tons trading from it, as well as many smaller boats. There were several new buildings such as the Northern Meeting Rooms and the Sheriff-Courthouse and fine Steeple, and facing them on the Castle Hill still stood the battered ruins of old Fort George. The centre of interest was still the River Ness, spanned by the seven-arched stone bridge. In a town which had hardly broken out of its medieval shell (none of its streets or houses was far from the River), the townsfolk were intrigued by the prospect of an artificial waterway to the west.

The same kind of activity was re-invigorating Lochaber. At Corpach, workshops, houses and turf huts were being erected and even a brewery was going up – an attempt to woo the Highlander from the 'pernicious habit of drinking whisky'. Cows were kept to supply them with milk and £400 was paid to Alexander McIntyre for oatmeal to be sold to the workmen at cost to make sure that basic foods were always available. In spite of the missionary spirit which led to the construction of the brewery, there were times when whisky was

Canal entrance at Corpach, light-house and lock-keepers' houses, and Ben Nevis in the background.

conceded to be so necessary that the authorities provided it. Whisky to the value of about £30 a year was allowed regularly to men who were working in the water until 1807 when their Lordships at the Treasury objected.

Two stone-carrying vessels were being supplied, another *Caledonia* (named after the Canal), and the *Corpach* being built at Chester, which was to bring up a steam engine. Rails were coming north by sea from North Wales and Memel timber from Greenock, while Colonel Cameron of Lochiel was having a great number of trees cut in his forests. This local birch was being sawn up at a sawmill at Strone, driven by water from the River Loy, into sizes suitable for making waggons, barrows and trestles. A rubble stone quarry was being worked at Fassifern and some good granite had been found near Ballachulish. Limestone was coming in from Sheep Island off Lismore and being burned in kilns just on the west side of the basin at Corpach. From a rubble quarry at Banavie a tramway had been laid to carry stone for building the first culvert.

Nearer the middle of the Glen too there was much activity. Colonel MacDonell of Glengarry set his men to tree-felling with enthusiasm and supplied timber worth £1,792 in a year. In the east, the Inverness New Foundry was started, providing iron rails in 1809 which were declared to be 'of an excellent mould and quality, for the maintenance and occasional extension of the railways near Clachnaharry'. It can be seen that all along the Glen, and particularly at each end, the Canal project was creating work, far in excess of the numbers directly employed on the Canal. The Canal workers themselves must have stimulated economic activity when they spent their monthly wages.

Telford expressed moderate satisfaction with the way the work was progressing when he reported in February 1804:

> The Basins at Corpach and Clachnaharry, which have been set out, have been carried on with as much success as could be expected in the winter season, and in a kind of work perfectly new in that part of the country:- the people have already fallen into the necessary modes of employment; and will soon, I have no doubt, acquire habits of industry.

Conscious of the scale of the work (cutting the line of the Canal in itself would require the digging and removal of five and a quarter million cubic yards of soil and rock) Telford saw that he would have to encourage and educate Highlanders to acquire those habits of industry. He did this partly by bringing organisers and pace-setters, 'those persons of experience in the several departments of Canal labour whom we have found it expedient to encourage to settle on the line of the Caledonian Canal, in order that they may undertake the contracts for work, and by their example impart skill and activity to the persons employed under their directions'. Secondly, it was decided early that as much work as possible would be done by measurement or piecework, 6d to be paid per cubic yard for earthworks, 2/- per cubic yard for cutting out rock in Corpach Basin and 11/- per cubic yard of rubble masonwork. 2/- per cubic foot was the rate for copings and common quoins using Ballachulish granite, 2/5d for cut freestone from the Cumbraes for hollow quoins to take the weight of the heel-posts of the lock-gates at Corpach, compared with 1/7d a cubic foot for similar work with Redcastle sandstone at Clachnaharry. These rates were lower than on some current works in England but the rate for day-labourers was fixed at 1/6d a day, the usual wage for labourers elsewhere in the Highlands. The refusal of John Telford to pay high wages at Corpach caused trouble there in October 1804, but as far as the Commissioners and their Engineer were concerned, it was quite clear in their minds that they were engaged in great public works, not relief work.

More men were offering themselves for employment. There were about a hundred and eighty in each basin by the summer of 1804 and soon cutting on the line of the Canal began. Thomas Davies started in September on a stretch at Kinmylies with ten men and by December he had one hundred and twenty-three. In all there were four hundred men at Clachnaharry by the end of the year, including over forty quarrymen and masons. Numbers employed at Corpach fell slightly as work on the basin could not proceed until a steam engine for pumping arrived. Simpson and Wilson, the masonry contractors, had turned to canal-cutting with a gang of between forty and fifty labourers on Corpach Moss, while other contractors, including John Meek, had smaller gangs at Banavie, Muirshearlich and Moy. Other labourers were floating timber along the Lochs. By the end of March there were four hundred men at Corpach including seventy-seven masons; by the end of July five hundred and forty-five men, including one hundred and seventy-three masons. At the same time, Clachnaharry had recruited over six hundred men, but fewer of them were masons. The build-up in the workforce had been rapid.

Where had the men come from? There seem to have been few masons in the Highlands. Davidson brought a number of Welsh masons and contractors with him from the Ellesmere Canal to Clachnaharry. An item in the Expense Accounts to May 1805 makes this quite clear:

> Sundry Expenses and Travelling Charges incurred by the above Persons [engineers including Telford and Jessop, John Howell, a map-maker and the superintendents] in the business of the Caledonian Canal; and also for Travelling Charges allowed to sundry Foremen and experienced Contractors, to enable them to remove their Residence to the Caledonian Canal £444.7.5d.

Not all his 'Welsh masons' were Welsh. Some, like Simpson and Wilson, were English, some Scots. The habit of the young mason

Clachnaharry village, hemmed in between the road and the Canal reach and split by the railway which passes over the Canal and under the road.

leaving home to make his fortune where there were public works was already established. Telford himself was an example of this. Others went off to work away from home for a season. As Telford warned the Commissioners in February 1804, 'Most of the most useful of them set off for England or the low parts of Scotland, early in the Spring'. In his *Reminiscences of My Life in the Highlands*, Joseph Mitchell recalled how the masons of Moray left home carrying their tools and little bundles of belongings for their work on the Caledonian Canal. Until the First World War, in fact, it was common for Scottish masons to go to America to work in the building season and return home with a money-belt full for the winter. Tradesmen came from Wales, from Moray and Nairn and were joined by men with experience of public works on the Forth and Clyde Canal. The labourers in the west came from overcrowded settlements in Argyll and west Inverness-shire including Skye, while those in the east came from Caithness, Ross-shire, Moray and even Aberdeenshire, many of them displaced by agricultural improvements.

Highland labourers had to learn to use the pick, the shovel and the wheelbarrow in canal-cutting conditions. After topsoil had been set aside, they dug down layer by layer, barrowing the soil along planks to the nearer side to form the upper part of the bank. The deeper cutting became, the steeper was their struggle up the planks, until it became necessary to bring in horses to help. Because the Canal bottom had to be level throughout the length of a reach, they did not dig so deeply in low land but had to build higher banks, using surplus soil from the nearest deep cutting. Labourers in deep cuttings pushed their barrowloads up on to a platform and emptied them into waggons below, to be conveyed along the railway to make up the banks on the lower ground.

Artificial banks were particularly porous and needed an inner lining of puddle. This mixture of light clay and sand or fine gravel had to be thoroughly mixed with water by agitating it with a spade until it became easily workable. It was applied in layers to a depth of two or three feet and worked in by 'clicking' or treading it with the feet. Special boots were used for this and they were expensive: an item in the accounts for 1820 reveals that four pairs cost £9. Puddle had to be protected against drying out in strong sunshine and loose earth was thrown on top of it to try to keep it moist. As puddling was best done in damp weather, the Canal labourers' work often changed with the seasons, cutting in summer and lining in winter. Puddling was slow and expensive and in the rainy Highlands Telford did not worry too much about the Canal losing some water. He did not intend to line the whole Canal but only stretches of loose gravelly soil and places where leaks would cause damage to nearby property. These two jobs, cutting and puddle-lining, were available to a large number of Highlanders, limited only by the amount of Government grant of £50,000 a year.

Having been warned, the Commissioners realised early that some labourers could not be relied on to work for the whole year, as they returned home at certain seasons to take part in peat-cutting, herring-fishing and harvesting. This was particularly true at Corpach. John Telford complained that some had not returned after the harvest because they were not in the habit of working in the winter. His feeling of irritation is understandable: it is difficult to make progress with an unreliable and half-trained labour force. What he did not realise, nor did Thomas Telford, was that the Highlanders were not

used to daily labour for someone else at all. They were used to working
to provide for their modest needs from their little patches of soil, their
grazings and their peat-banks. They needed some work to supplement
their traditional way of making a living, not a complete substitute to
make them want to give it up. Work for a wage was a novelty, all right
now and then, but to do it six days a week, every week?

On 21 October 1805, Lord Nelson died, knowing that he had
destroyed the French and Spanish fleets off Cape Trafalgar. On
22 October, having heard nothing about Lord Nelson's famous victory,
Matthew Davidson at Clachnaharry and John Telford at Corpach
were completing their monthly labour returns. Davidson's workforce
was five hundred and thirty-one men, nineteen more than he had
had in September. The biggest number were at the sea-lock, forty
masons and one hundred and seven labourers: he had two big cutting
contracts going, one employing one hundred and thirty men making
up the banks through the low land at Kinmylies, the other with
eighty-four men at Holm and seventy-eight at Dunaincroy making
embankments to change the channel of the River Ness; there were
twenty-two masons and twenty-four labourers cutting and loading
stone across the Firth at Redcastle and seven sailors manning the
boats. That was the state of his campaign.

At Corpach, John Telford had some reason for gloom. He had only
half the labourers that he had had in August, worse still, only a third
of the number he had had in July. Even the number of masons was
down by a third on August. Still, the engine house was now up at
the sea-lock and the engine was working, and fifty-four masons and
twenty-eight labourers were working there and at Banavie Aqueduct.
Only twenty-six labourers were now at Corpach Moss with a mile
completed, thirty at Moy with half a mile completed, thirty at Banavie
with half a mile completed, only eight men left in the deep cutting at
Muirshearlich and they were hardly at the bottom anywhere; twenty-
seven masons in the three quarries, Fassifern, on Lismore and the
Cumbraes; six timber floaters – a total of two hundred and forty-one.
Oh, to change places with Davidson in Inverness!

Superintendents did not have to deal with landowners unless, as
sometimes happened, landowners objected to work being started on
their land before it had been paid for. Usually, Thomas Telford and
George Brown, the Elgin surveyor, walked over and measured the

land needed for the Canal and Brown valued it. For example, on 3 September 1805, Brown noted in his Journal, 'Left home along with Mr Telford and came to Inverness to settle land to be occupied by the Canal'. The plans he made still exist, in the care of the National Archives of Scotland. His valuation was the basis of the price offered to the landowners. If they refused it, the matter was submitted to a jury. The following month, he recorded his feelings that juries in Inverness had been more generous than his professional judgement allowed him to be. About the compensation for the lands of Muirtown he wrote, 'He got too much by £400', for Kinmylies 'a little more than my valuation', Dochgarroch '£90 more' and Dunain 'a little more'.

All things considered, what had been achieved was commendable. A huge workforce had been recruited and organised, the materials for construction had been assembled and work was well under way. Immediate problems had been solved, and many long-term difficulties identified, thanks to the professional knowledge and experience brought north by Jessop and Telford.

6
EAST END

Matthew Davidson knew the plan for the whole Canal in outline and as resident superintendent at Clachnaharry he knew that part of it on his doorstep better than anyone else. A hard, capable, dependable man, with a cynical wit, he was much addicted to literature, bathing in cold water, and cursing Scotsmen. John Telford, his opposite number in Corpach, maintained that 'Davidson would not accept a seat in Heaven if there were a Scotsman admitted to it': Davidson seems to have forgotten that he was born in Dumfriesshire!

When he looked out from his new house at Clachnaharry (beside the old Canal Office) in 1807 he saw the Canal works before him. Living beside his work was nothing new to Davidson for he had lived beside the Pontcysyllte Aqueduct for eight years, presiding over its progress. Across the road to the left he saw the timber yard and workshops for carpenters and blacksmiths and behind them the stone-lined U-shape of Clachnaharry Lock, the first to be completed at his end of the works. It had been skilfully built by Simpson and Cargill and their men and, judged by his own high standards, it was a good piece of workmanship.

Away to his right, he had the satisfaction of knowing that the bottom of the great basin they were constructing at Muirtown was above low-water mark. A dyke had been built across the old sea inlet to protect the work against high tides. In 1806 when Telford was up on his spring inspection half of it had been excavated to the required depth and the soil used to build up the banks. By September the next year, it would be finished – no mean feat when it is considered that it measures 800 yards by 140. When the wharf was built at the east end, it would share in the trade of Inverness. Along the road towards Inverness, now called Telford Street, Simpson and Cargill, his masonry contractors, had built a new block of houses of red sandstone

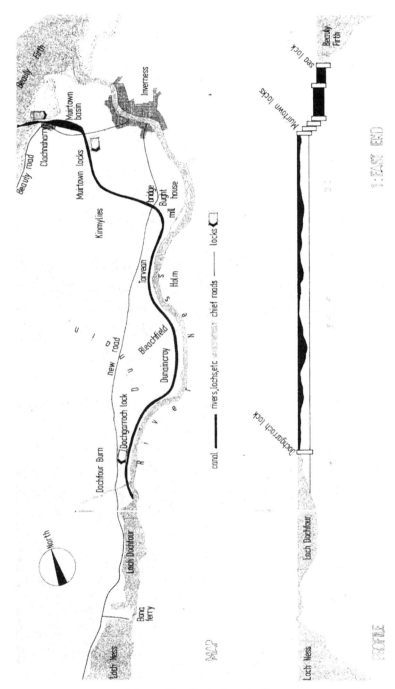

Map 1. East End. *Douglas Stuart*

Lock-keepers' view today over the Clachnaharry Works Lock, the first lock to be completed on the Canal, and now with a bridge carrying the railway across it.

for themselves and other contractors and overseers. John Mitchell, Telford's Chief Inspector of Roads after 1814, also lived there. The block still stands between the junctions of Telford Road and Lochalsh Road with Telford Street.

To the left, out of sight from his house, was Davidson's main concern, the approach to the sea-lock. The two sea-banks he was having built there were thrusting their way out into the sea, fed by two 3 ft 3 in. gauge tramways, one to bring rubble and rubbish from the quarry, the other carrying earth under the Beauly–Inverness road from the hill behind. The great cavity gouged out of the hill next to the old Canal Office is still known locally as 'the Hole'. A year ago, one mound was 240 yards out from high-water mark – two-thirds of the way there. Now in the spring of 1807 the west bank was far enough out, the east one more than half way there. Davidson reckoned it would be another two and a half years before the sea-lock was built. When the sea-banks were ready for a coffer-dam to be erected to hold out the sea while the sea-lock was constructed between them, work had to be postponed because the price of the foreign timber necessary had risen so much.

Cold wintry weather at the beginning of 1808 slowed the work. Davidson had only eleven masons working in January and recorded none at all in February. The masons who had worked with him in Wales were probably still at home. It was not till April that he had fifty masons at work again, and that month there were eleven days of snow or hail or sleet. Matthew Davidson probably cursed and indulged himself with a bathe in the Firth.

Work was then concentrated on the four locks at Muirtown, the lowest finished, the second lowest half finished, and the upper locks being excavated. Stone was being delivered to them regularly from the pier and the quarry at Clachnaharry in special waggons pulled by horses along tramways a mile long. The masonry of the four was completed in 1809. Above the locks the cutting was almost finished through Kinmylies and the loose gravelly soil of Bught and close to the bottom of the steep Torvean hill. Across the river, land was cut away in Holm estate in order to move the bed of the River Ness further to the south-east to allow the Canal to occupy part of the river bed at the base of Torvean hill. This required the raising of a massive embankment, a thousand yards long to support the Canal well above river level and to continue the intended towpath on the south bank along the Canal. During the year 1808 soil cut from the base of Torvean was carried by tramway to make up banks of the Canal in the low-lying lands of Bught to the proper height. The contractors were Thomas Davies and William Hughes.

Much of Matthew Davidson's time was spent in setting out and measuring work, because so much was done at piece-work rates. As canal-cutting extended further along the river his journeys on horse-back to inspect the works became longer and more time-consuming. He was intrigued when some workmen, cutting through a cairn east of Torvean in 1807, found a massive chain with double links, all of eighteen inches long and weighing seven pounds of solid silver. The chain is now in the Royal Museum of Scotland in Edinburgh. But his mind was often on the problem of the sea-lock. He noticed that the sea-banks were settling and sinking into the mud. This gave someone – whether it was Davidson himself, or Telford or Jessop is not clear – the idea of dumping soil and quarry rubbish between the two banks, and more on top, to form a peninsula, then to let it settle and excavate a lock out of it, without the expensive alternative of making a coffer-dam.

By the end of 1809 the peninsula was complete and higher than ever. It was given another six months to settle and everything was prepared for efficient work below the level of the sea. During excavation below high-water mark, the sea came in. Hand pumping was tried, then a chain pump worked by six horses, but in vain. Recourse had to be had to a six horse-power Boulton & Watt steam engine, which had been unused in Clachnaharry since 1805, and first of all an engine house had to be built. The engine was powerful enough to keep the lock-pit dry. The lock was excavated eight feet into the compressed clay bed,

Houses in Telford Street, Inverness, built by Simpson and Cargill for the Canal contractors to live in. John Mitchell, Telford's superintendent of roads, also lived there.

which proved to be impervious to water. Foundations of rubble-stone were laid, two feet thick in the centre, increasing in a gradual curve to five feet at the sides. Complete rubble side-walls were erected to prevent the soil banks falling into the lock, and puddle was worked in behind the walls to seal them against leaks. Then the inverted arch bottom to the lock was put in by the masons and the high side-walls raised, all built of dressed Redcastle stone. Great care was taken with the hollow quoins against which the lock-gates would be anchored, the recess walls to take the opened gates, and the wing walls at the entrance to the lock. They were all built on a foundation of timber, pile-driven into the mud. So this unique sea-lock was finished in 1812 and a victory won against the sea, but three years later than Davidson had predicted.

Meanwhile the new position of the Canal cutting so close to the east of Torvean hill destroyed part of the old road to Fort Augustus. Cutting into the base of so steep a hill to replace it would not only be laborious

By Torvean, the Canal occupies the old bed of the River Ness and has been raised above it, protected by this artificial bank where broom and whin still grow, descended from the seed Jessop and Telford ordered to be sown there.

but dangerous because of falling rock. It was decided therefore to build a better road, the present road, on the other side of the hill from Tomnahurich Bridge for three and a half miles to Dochfour Burn. It was constructed in a year under the supervision of an experienced road-maker, Donald McKintosh, for the sum of £1,172.

Contractors for canal-cutting pressed on, through the boggy land at Dunain, and westwards turning an island into a bank between Canal and river, and in places moving the bed of the Ness eastwards to let the Canal take part of the old river bed. At Dochgarroch a suitable site for a regulating lock was found. Regulating, or guard, locks were necessary at the ends of the Lochs to guard the Canal against a sudden rise in the level of water in the Lochs. Beside the Dochfour Burn a tramway was laid down for a length of half a mile to carry stone from a rubble quarry for the lock and the weir to be built later at the east end of Loch Dochfour.

Good timber was scarce and becoming very expensive. The demand from the Navy for building and repairing warships was heavy and supplies from the Baltic had dried up. This caused delays and changes of plan. The construction of the 150-ton barge of oak and pine to carry the steam-powered dredging machine on Loch Dochfour had to be postponed. Not until 1814 were this Butterley dredger, the *Prince Regent*, and its six punts at work making the channel into Loch Dochfour deep enough to take vessels with a draught of ten feet.

By that time gates of Welsh oak had been made to William Jessop's design by the carpenters and smiths at Clachnaharry and hung at the sea-lock and second lock. For lock-gates which would not come in contact with salt water, and for bridges as well, Jessop and Telford decided to switch to cast-iron in 1810, because it was of British manufacture and easy to obtain. Telford declared his faith in cast-iron, 'It will in future prove as serviceable in large lock- and dock-gates here and elsewhere, as I have already in many years experience found it to be in gates of smaller dimensions, as well as in large bridges and aqueducts'. This also reveals that the lock-gates of that material on the Caledonian Canal were the biggest he had ever handled. The cast-iron for lock-gates for the east end was ordered from Butterley Iron Works in Derbyshire, for the west from Hazledine's at Plas Kynaston in Wales, and to be delivered by sea.

July 1815 brought news of the Duke of Wellington's victory over the French at Waterloo. The war against France was at an end but

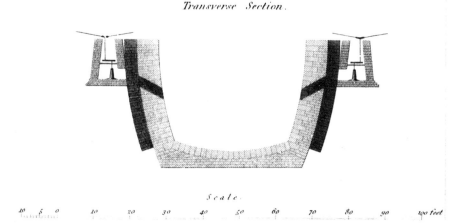

Transverse Section.

Scale.

Section of the sea-lock at Clachnaharry showing winding gear for operating the gates.

the Canal, intended to be useful for naval frigates in wartime, was still far from ready. It was not even navigable between Inverness and Loch Ness. Telford's original estimate that the whole Canal could be completed in seven years was shown to be wildly optimistic.

In the following year, the cast-iron lock-gates at Muirtown and Dochgarroch and the horizontal turn-bridges, recommended by William Jessop, at Muirtown and Tomnahurich were installed. Nearer Loch Ness, pools and waterfalls were formed in the Dochfour Burn to prevent so much debris being deposited in the Canal when the stream was in flood. Meanwhile, the dredging operation in Loch Dochfour was proving successful. As the crews became more experienced under William Hughes' direction, the amount of material they could remove in a day rose to four hundred tons. The channel into Loch Ness was now fourteen feet deep.

At last, the Canal was ready to be tested by letting water flow in from the River Ness to a level of eight feet. Most of it was watertight but it was the leaks that people noticed, especially the two families at Bught Mill whose cottages were 'rendered uninhabitable', and Provost Grant of Bught House, who, finding his cellar flooded, expressed considerable annoyance. Only the artificial banks of the Canal had been lined with clay, not the natural banks or the bottom, in the belief

Tugs moored in Muirtown basin: it was hoped at the time of its construction that it would become a second port for Inverness.

that sediment would soon make it watertight. The Canal was emptied and for a length of a mile through the lands of Bught, the bottom and banks were lined with clay puddle – 'a tedious and expensive process', as the Commissioners admitted.

The Canal through to Loch Ness was opened to traffic during the summer of 1818 and was used by about a hundred and fifty coasters bringing in oatmeal, coals, tar and lime to Fort Augustus and carrying out timber and wool. More important for the progress of the Canal, building materials could be brought up from Inverness more easily and cheaply.

Matthew Davidson died in February 1819, having seen the Canal opened to Loch Ness. In his earlier years in Clachnaharry he had turned an open shore into the entrance to a great waterway, its locks and basins. Probably the climax of his achievement in the Highlands was the sea-lock built in the mud at Clachnaharry. It was certainly the climax of Jessop's, for he signed his last joint report with Telford in October 1812 and died of paralysis soon afterwards. The quality of the masonwork at Clachnaharry and Muirtown is a tribute to Davidson's keen eye and his insistence on high standards. He saw the Clachnaharry quarry closed down and the tramways lifted for work further west and he organised the great leapfrogging movement of men, materials and work to Fort Augustus and beyond. Delays over lock-gates and dredging dimmed his satisfaction in the years after 1812, for he was a man 'zealous to a degree of anxiety', to quote the Commissioners' tribute to him. His son James, who had taken his place during his illness, succeeded him as superintendent and became responsible for the completion of the Canal operation from the east.

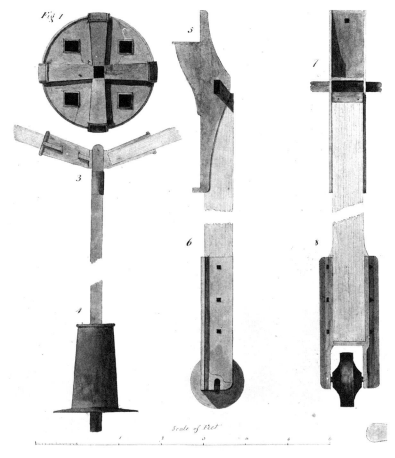

Ironwork for the locks, drawn by Thomas Rhodes, the carpenter from Hull who built the gates.

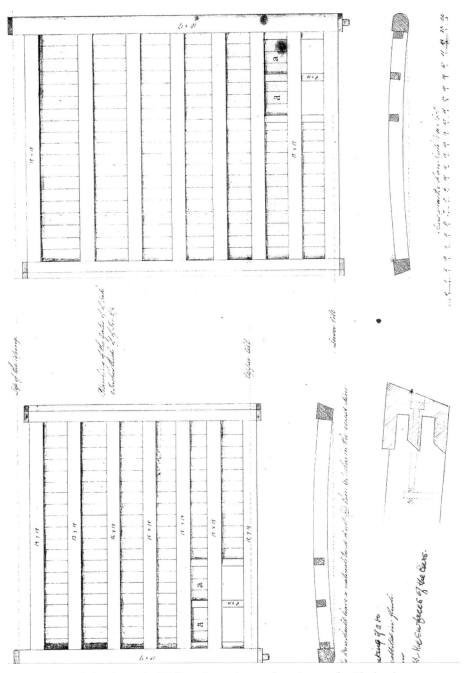

Original plan drawn by William Jessop in 1810 for oak gates for Clachnaharry sea-lock and rediscovered in 1992; it allows for 2½ in. thick fir planking to be used if 2 in. oak cannot be obtained.

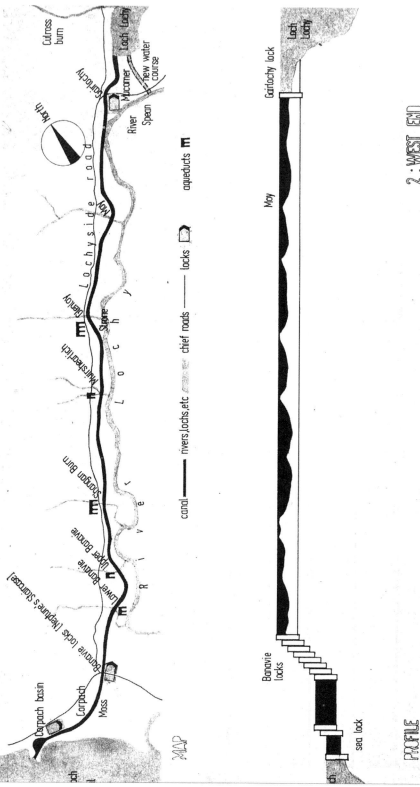

MAP

Culross burn
Loch Lochy
Mucomer
new water course
River Spean
Mucomer
North
road
Lochyside
Glenloy
Stone
Muirshearlich
Loch
Spean burn
y
River
Upper Banavie
Lower Banavie
Banavie Locks [Neptune's Staircase]
Corpach basin
Corpach
Moss

canal ——— rivers,lochs,etc ≈≈≈ chief roads ——— locks ⌂ aqueducts ᴟ

PROFILE

Loch Lochy
Gairlochy lock
Moy
Banavie locks
sea lock

2 : WEST END

7
WEST END

John Telford's first concern at Corpach, like Davidson's at Clachnaharry, was to secure an entrance from the sea. The point chosen was between Corpach House and Kilmally Kirk on Loch Eil, where the water was deepest nearest to the shore and sheltered from the prevailing wind by three small islands. This place was about a mile west of the mouth of River Lochy, on the opposite side from Fort William. Like Davidson's, his men would have to work below the level of the sea and a steam engine had to be ordered before they could start. His construction problem, however, was quite different because the entrance and basin at Corpach had to be cut out of solid rock. Work began therefore at the inland end of the basin, preparing the foundations of two locks, while at the sea end an embankment, being erected to surround the intended sea-lock, served as a quay for landing materials coming in by sea.

Because of the delay while the 20 hp steam engine was being constructed by Boulton & Watt in Birmingham and because so many men were applying for employment, John Telford had to adapt his programme of work. Rock-blasting and cutting teams began to shape a basin with regular sides and on the south-east side an engine-house was built to receive the pumping engine. Otherwise there was little building work meantime for masons at Corpach, now that the original buildings – houses, workshops, kilns and the brewery – had been completed. They could not start on the great flight of locks at Banavie until steps had been taken to bring in the materials they needed. Most of the masons therefore were set to work building aqueducts which would carry the Canal over the mountain streams to the east and many of the labourers were soon busy cutting sections of the Canal.

Opposite: Map 2. West End. *Douglas Stuart*

Glen Loy Aqueduct, hardly noticed by road travellers today, carries the Canal over the River Loy, with a 25-ft arch to accommodate the river in spate, and side arches for farm traffic.

William Jessop showed John Telford how to mark out the cuttings through level or sloping ground. From Corpach to Banavie the plan was to cut the Canal to half its depth and to use all the excavated soil to build up the south bank first of all, to allow a tramway to be laid along it to start transporting stone and lime to Banavie for lock-building to begin. All the time there were men preparing stone in the rubble quarries west of Corpach but good enough freestone was not to be found in the neighbourhood. It had to be brought by sea and through the Crinan Canal, in the sloops *Corpach* and *Caledonia* from the Cumbraes in the Firth of Clyde. A regular force of quarrymen was employed there and every ton was subject to a royalty of sixpence to the Earl of Glasgow. For these reasons freestone for the west was expensive compared with the Redcastle supply to Clachnaharry. John Telford's workforce was widely scattered but it was tactically deployed to prepare for the earliest possible start on the great masonry works which had to be undertaken.

He looked forward to Thomas Telford's two visits each year and Jessop's in the autumn as much for links with the outside world as for guidance in forward planning, and he wrote teasingly to Davidson envying him in the urban civilisation of Inverness. The works he had created at Corpach were a big enough construction job on their own by man's standards but he was conscious of their smallness in the vastness of the mountains, as he looked east towards Ben Nevis often shrouded in mist or on clearer days revealing its snowy head to the skies.

He did not need to contemplate his isolation: it was often forced upon him by incidents such as Highlanders demanding higher wages, or worse, in August 1805, the danger that he might not have enough money to pay the men. He told Davidson of his position, 'If all the men are not settled within Monday night at furthest, I dread the consequences. We are all well here at present; God knows how long we shall remain so.' He lacked the experience and, perhaps, the hardness of Davidson but there is no doubt that his position was more difficult.

Far more than Davidson, John Telford had to cope with large numbers who had no experience of working for pay. Some Highlanders must have looked at the rock-blasting and cutting in Corpach Basin with dismay and decided that it was not for them. Others took to that work. Many more were engaged under Simpson and Wilson on Corpach Moss, others above Banavie with Meek or Gillies and Ross. Thomas Telford was impressed that so many soon became skilful, hard workers but some men were found unfit for the strain of canal work. There was alternative employment for them on the expanding programme of roadworks, which suited many fit men too, if these roadworks were going on nearer their homes. It is not clear how great the turnover of workers on the Canal was, but Thomas Telford noted with pleasure in 1807 that even although some went off because of the seasonal call of the peats or the potatoes, many would return immediately if emergency required it. This would seem to indicate that the Canal soon had a solid core of reliable workers, Highlanders as well as imported pace-setters, labourers as well as masons.

Fortunately there were no difficulties over the purchase of land in the west. All the land between Corpach and Loch Lochy belonged to Colonel Donald Cameron of Lochiel. He allowed cutting to start right away and, without recourse to a jury, he agreed to accept the valuation placed on his land – just over £2,000 for one hundred and ninety-four Scotch acres, the equivalent of thirty years' annual rental.

At Lower Banavie the first of the culverts under the Canal was reported complete in May 1806. It had two arches, nine feet wide and ten feet high, one to allow the burn to pass under the Canal, the other as a through-way for farm carts and cattle. The scale of the Canal was becoming clear from the length of these passages under it: they were eighty-four yards long. A single-arch aqueduct over the burn at Muirshearlich was under construction and near it the Canal was being cut through the edge of a hill. This required much deeper cutting and produced far more soil than was needed for the banks there. The intention was to complete that section of the Canal, fill it up with water and transport the surplus soil in boats to make up the banks on the lower lands at Strone to the north-east. Strone was a centre of much early activity. There was a sawmill there, worked by a mill-dam from the River Loy, and beside it a carpenter's shop and a timber yard. The Glen Loy Aqueduct there was the biggest one to be constructed,

The road under the Shangan Burn Aqueduct is used by cars today far more often than cows.

The west end of the Great Glen, Banavie Locks (Neptune's Staircase) and quarry, Canal and River Lochy side by side and Loch Lochy in the distance.

having a central arch twenty-five feet wide to accommodate the river in flood and two ten feet side arches for farm traffic. Simpson and Wilson concentrated nearly a hundred masons on it in the summer of 1806 and finished it by October. Timber for battens and trestles, and to support the arches was produced on the spot and rubble stone was found not far away.

Between these two, work began on a single arch at Upper Banavie and a three-arch aqueduct over the Shangan Burn. Nearer home, the two locks above Corpach Basin, built together to save one lock-gate, were now complete but John Telford was anxious to see work begin on the sea-entrance and the lock-ladder at Banavie. Neither of these was to be his achievement nor his worry, for he died in April 1807.

The wife of this 'diligent and faithful servant' had no reason or wish to stay among the mountains of Lochaber and returned to live in Chester, aided by a gratuity of £50 but no pension. John Telford's place at Corpach was taken by Alexander Easton who had been a mason on the Forth and Clyde Canal and had proved himself as an inspector of roadworks in Argyll. Easton's promotion was an example of Thomas Telford's willingness to promote the practical man who had shown strength of character.

Easton saw the remaining aqueducts completed in his first season and the idea of an aqueduct for the Moy Burn, further east, abandoned, because the foundation would be below the level of the nearby River Lochy. Instead it was decided by Jessop and Telford to allow this stream to flow into the Canal and construct an overflow on the other side. To control the amount of debris which might be deposited in the Canal, the Burn was to be 'disciplined' by forming a series of waterfalls and pools in it to trap stones and gravel brought down by floods. The five-arch inlet for the Burn to enter the Canal was built in 1813 and the outlet the following year.

The problem at Banavie was to raise ships from the level of Corpach Moss steeply to a much higher level, over sixty feet higher, within a space of about five hundred yards. Locks had to be built in a flight or ladder formation, with no less than eight locks in a row. Besides being built in a shorter time, this led to considerable savings in materials, especially with regard to lock-gates. Only nine pairs of gates were needed compared with sixteen for eight single locks. When the Canal was in operation, the disadvantage of this arrangement became evident to every skipper who had to wait until another vessel in transit in the opposite direction had completed its journey through eight locks before he could enter.

Since work on these locks was expected to last four years, it was sensible to start by building two substantial houses for lock-keepers, where some of Simpson and Wilson's masons could live while they were employed there. The horses were stabled at Corpach and were used to bring loads of stone and lime along the one and a half miles of new tramway. Each horse pulled a three ton load of stone in a waggon up an incline of ¼ inch per yard. Working a ten-hour day, each covered fifteen to eighteen miles every day.

The first lock was excavated during Easton's first summer as superintendent. It was not till April after the cold winter of 1808 that the masons returned in strength but Simpson and Wilson employed

Empty locks at Banavie in 1920

every available mason at Banavie. An average of nearly a hundred and fifty men worked all through the building season to late November and had nearly three completed locks to their credit. The next season there were fewer masons but still more than a hundred, backed up by gangs of quarrymen at Fassifern and in the Cumbraes keeping up the supply of stone. The number of completed locks rose to five. If the masons went home in the winter months, the men in the quarries did not. Throughout the 1810 season the building work was carried on with vigour by over one hundred and twenty masons with another sixty in the quarries at Banavie, Fassifern and the Cumbraes.

The Commissioners noted with satisfaction in May 1811 that the masonry of the eight locks was finished, except for the freestone coping of the last, in all, five hundred yards of solid masonry. Telford, a former mason himself, expressed his belief that 'there does not seem to be the slightest imperfection in this immense mass of building'. He had estimated that the work would take four years: it had been completed in three and a half. This was adjudged to be a considerable achievement, owing much to good organisation, adequately prepared stocks of stone, good though long lines of communication and a large labour-force; but the completion of works on the Canal within the estimated time was unusual and perhaps Telford should have tempered his admiration with a dash of suspicion.

Back at Corpach the formation of the Basin out of solid rock was proceeding more slowly. Not as big as Muirtown Basin in the east, its dimensions (250 yards in length by about 100 in breadth) were still daunting in relation to the tools employed and the hostility of the sea. Down to a safe level the rock was excavated and built into embankments against the sea. The engine was installed in the engine-house, the well for the pumps was sunk into rock and connected with the excavation by the beginning of 1810, and arrangements were made for a skilled engineer to come to Corpach 'to instruct our Superintendent in the perfect use of the Engine'. The coffer-dam, which was necessary to hold out the sea while the entrance and sea-lock were under construction, was very difficult to build. Many of the main piles could not be kept upright until ironwork was driven into the bedrock and fixed to them to hold them in position. Close

Opposite: Tall ships at Banavie in July 1991.

piling formed the outer defence against the sea. It was backed by inner piling and framed bracing. Storms interrupted and destroyed parts of the unfinished work and delayed its repair. By the summer of 1810, however, it was ready.

Easton put the steam engine in motion, excavation began and went on through the winter without any lessening of effort. Behind the dam, Simpson's and Wilson's men were gradually lowering the level through solid rock until the required depth was reached for the whole Basin. It was a long and tedious occupation – blasting, cutting, shovelling, barrowing, day after day, week after week. The men who had been building the locks at Banavie, 'Neptune's Staircase' as they now called it, were available in strength the next season for work on the sea-lock. For four months over two hundred masons were building there at full stretch, because the steam engine had to be in operation pumping out water for two hours out of every four until the works were complete and the lock-gates hung. When Jessop and Telford were there in October 1811 the carpenters under James Rhodes were making the oak gates which were in position the following spring. The sea-lock at Corpach was the first lock on the Canal to be made ready for use, because the engine could not stop pumping and the coffer-dam could not be removed until it took over the task of holding out the sea. It was an unusual coincidence that the sea-locks at Corpach and Clachnaharry, each so different and each delayed for different reasons, should be completed at almost the same time.

The operations of Simpson and Wilson as contractors were expanding. Even when they had the big masonry contract for Banavie Locks, they undertook not only to make the road from Banavie along the north-west of the Canal as far as the Culross Burn flowing into Loch Lochy, but also to build the bridges beyond the Burn to the north of it on the stretch of new road being constructed by men under Colonel Cameron of Lochiel. The road contract, signed at Corpach in 1809 and witnessed by John Duncombe and Alexander Easton, is of great interest. It required the building of no less than fifty-six one-arched bridges in a distance of about ten miles and goes a long way to explain how it was possible for Thomas Telford to be credited with building 'over eleven hundred bridges in the Highlands'. Neither Telford, nor Duncombe, nor Simpson nor Wilson drew any plans for them. Telford simply laid down general rules about the foundations

'Neptune's Staircase' as it used to be with the sluice mechanism on top of the gates.

and the height and width of arches, and the masons built them. The road was to be sixteen feet wide, with fourteen inches of gravel in the centre reducing to nine inches at each side. Among the pages of legal language the kind of gravel was specified with a down-to-earth touch: it was to be 'of a proper quality . . . out of which all stones above the size of a Hen's Egg shall be taken'. His road inspectors all carried ring gauges to check it! Probably the real reason for Simpson and Wilson taking the contract was to keep their men employed. Rises in the amount spent on timber that year meant that less of the annual grant of £50,000 for the Canal works was left to pay labourers, and these roadworks were charged equally to the Roads and Bridges Commission and the neighbouring proprietor. The road was ready ahead of schedule in the summer of 1810.

Roadmaking showed results far more quickly than canal-cutting. The farther inland the Canal labourers moved, the more difficult it was to get supplies of food, especially oatmeal. Food shortages meant fewer workers and slowed down the work. But the main barrier to

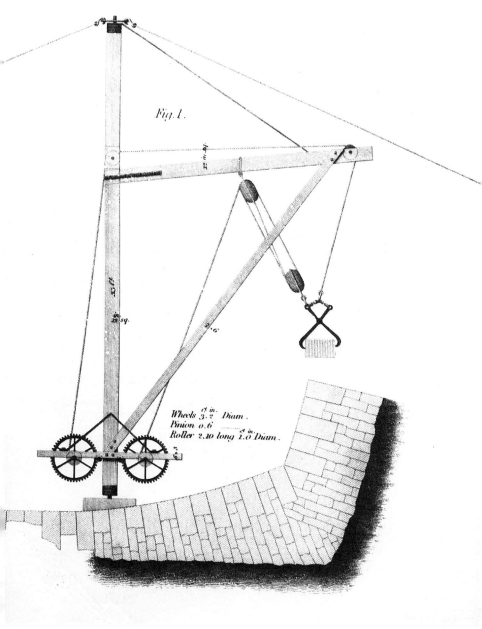

Fig. 1.

Wheels 3.² Diam.
Pinion 0.6
Roller 2.10 long 1.0 Diam.

Crane used in the building of the locks.

progress was the difficulty of moving large quantities of earth from deep cuttings to make up the banks on lower ground. The earlier idea that a section of the Canal could be filled with water and the soil transported in punts had to be discarded. Railways were the only answer but the rails were still needed between Corpach and Banavie. At length a railway was laid along the north-west bank from the Muirshearlich cutting and earth moved out of it in both directions in horse-drawn waggons. John Meek, a good organiser, commanded a huge army of workers, over two hundred and fifty in 1811, and eighteen horses working on the railway, but that was an exceptional year. In 1812 he had fewer workers and several of his horses died. Fodder for them was scarce and expensive and he tried the experiment of replacing them with oxen. He told Telford he was very impressed with them because they stood up to the work better than horses. Rising costs meant that Meek was making losses on his contract at prices which had seemed reasonable earlier. Telford examined the position carefully and increased his rates, including an extra halfpenny for every cubic yard he had moved already. John Meek sold his oxen and bought thirty horses!

As the years passed, the work slowly progressed. In their 1814 Report, the Commissioners predicted, 'A quarter of a mile of bank remains unfinished, and at the present rate of progress will not be completed in less than two years' – and they were right. This difficult section showed the limitations of men with picks and shovels and horses pulling waggons when soil had to be moved a long distance.

The only suitable entry for the Canal into Loch Lochy from the south-west was in the bed of the River Lochy. Jessop and Telford planned to cut a new channel for the River for half a mile to Mucomer to lead its waters from the Loch into the River Spean, which in turn joined the River Lochy farther down. The regulating lock was to be constructed on solid rock at Gairlochy. Worried about the cost of bringing freestone from the Cumbraes to Corpach by sea and inland to Gairlochy by tramway, Telford explored the shores of Loch Lochy and found serviceable rubble near Clunes four miles away. Preparations were made quickly. A 40-ton barge was constructed and a railway laid to the lock-site and, as had been done at Banavie, the lock-keeper's house was built first as a lodging for the workmen. The middle room upstairs with a wide view from its bow window is still known locally as 'Telford's Room', because he occupied it during his inspection visits. The masons built the lock in the seasons of 1811 and 1812 and exhausted the quarry's supply of stone. Then over the new channel at Mucomer, an important route for drovers bringing cattle from Skye, they constructed a fine and substantial bridge which is still in use. Beneath the bridge they quarried the rock bed to give the new River Lochy a splendid entry over a waterfall into the Spean.

Although all the construction works to the west of Loch Lochy were ready by 1816 they could not be put into operation because the level of the Loch had to be raised twelve feet and that could not be done until the works in the centre allowed it. Like Davidson in the east, Alexander Easton was now seeing the main operations in his area being performed farther away from his base. Except for the carpenters and smiths, there was no one now at Corpach. Except for Meek's men,

Telford's splendid Mucomer Bridge over the new water course he made for the River Lochy: picture taken during preparatory work for the Mucomer Power Station.

still finishing off their deep cutting, there were few on the western line of cutting. The bulk of the labourers were now at Laggan at the north-east end of Loch Lochy, some in their third season there.

It would have been surprising if such a large public undertaking could have gone on for so many years without some labour troubles. Earlier, John Telford had encountered some difficulties about pay rates and the availability of money to pay the men. In 1812, Alexander Easton was accused of collusion in fraudulent measurement of work by a former labourer, William Grant, then living at Camlachie, near Glasgow. An enquiry was held by magistrates at Corpach, with Jessop in attendance to advise on technical aspects. None of the accuser's witnesses confirmed his assertions. Instead they maintained that Easton was too strict in measuring work and the result was that he was completely cleared. Grant kept writing to workmen at the Muirshearlich cutting and between them they presented a memorial containing fresh complaints, signed by John MacLennan, Angus McNaughton, Angus McDonald, Charles Cameron, Dugald McPhee, John Cameron and John McMaster. They added the name of Duncan MacDonald falsely. It is only on occasions like this that the names of any of the labourers on the Canal are known to us. Their claims were considered and refuted. Easton and Meek were given permission to sack those who signed the document and were instructed to tell all the other men that they need not continue to work there if they did not wish to.

The influx of workers to the west end of the Canal was at its peak at this time. Napoleon's Continental System, an economic blockade to keep British goods out of Europe, was having severe effects, causing wage cuts and unemployment, especially in industrial centres like Glasgow. In addition to weavers and factory workers, hundreds of masons and labourers were out of work. Many of them made their way to Corpach, where in August 1811 there was almost a thousand workers, including three hundred and sixty-six masons. This level of employment could not be afforded out of the annual grant for long but the Canal works, started to check emigration from the Highlands, also provided some temporary work for men thrown out of work in Glasgow and district.

8

IN THE CENTRE

From an engineering point of view, it would have been advantageous to have the Canal open from the east before work began in earnest at Fort Augustus but the opening was delayed till 1818. Thomas Davies and William Hughes, the contractors who had been responsible for most of the cutting at the east end, had men to spare in 1811, the year when the number of men engaged on the Canal was at its peak, and set them on excavation in the centre. In the early years some preliminary work had been done at Fort Augustus but difficulties in finding a suitable site for the flight of locks had brought these operations to an end in 1806. That decision was adhered to, in spite of a petition signed by over three hundred people pleading for work at Fort Augustus to be resumed.

Fort Augustus, where the Government fort had given its name to the village formerly called Kilcumein, was still occupied by military veterans, on the site of the later Benedictine Abbey. Labourers flooded into the village, half of them getting lodgings in the village and farms round about, the rest living in huts provided by the contractors. The huge stock of picks and shovels, barrows and planks brought in to start the works proved such a temptation in an isolated community that a great deal of pilfering took place. Being unwilling to establish elaborate stock-checking and extra supervision, Telford passed the responsibility to the contractors. They had to supply their own planks and barrows and to build their own huts. In compensation, they were paid an extra ¾d per cubic yard of excavated soil.

Middle East – Fort Augustus to Loch Oich

The River Oich flowed north-eastwards out of Loch Oich for a distance of six miles through lands belonging to MacDonell of Glengarry and

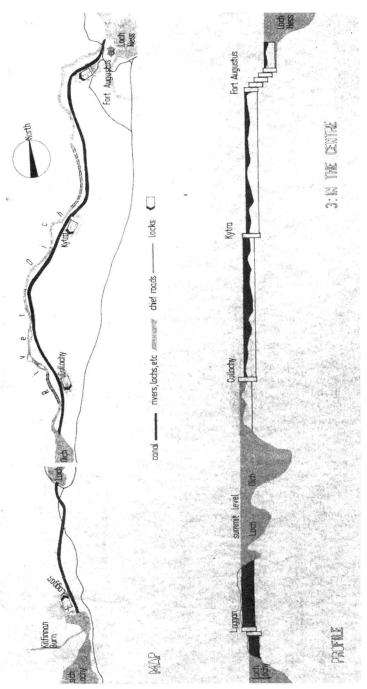

Map 3. In the centre, *Douglas Stuart*

Fraser of Lovat to enter Loch Ness at Fort Augustus. Jessop and Telford planned to cut the Canal through the low land close to the River, and on the south-east side of it, the opposite side from the cuttings at the east and west ends of the Canal. To give the Canal a straighter line and to avoid a lot of deep cutting, the River Oich was to be diverted into a new channel for nearly a mile and the river bed used for that part of the Canal. The soil from the new river channel had all to be wheeled to one side, the south-east side, to form a bank to protect the Canal. Not only was this slower because it required far more barrowing of soil by Thomas Davies' men, it was also more costly, 2d per cubic yard more than the usual rate. The River was diverted into its new bed in 1813. Nearer to Loch Oich, Hughes' men were excavating the cutting through Cullochy Moor and between them the two gangs had completed four miles of the Canal in this area by 1815.

It was the siting of locks that gave most trouble. After many trials a bed of rock was found crossing the line of the Canal at the correct depth and this position was chosen for Kytra Lock. Building stone was not imported from the east to build a lock in such an inaccessible place. Because of the rock foundation, an inverted arch of masonry was thought to be unnecessary. Granite was found on the Moor only three hundred yards away and a tramroad was laid down between the quarry and the lock. Oddly enough, the supply of stone was just sufficient to finish the lock and no more.

The cutting of the Canal on Glengarry's land nearer to Loch Oich was delayed year after year. George Brown, the Elgin surveyor, had walked over it with Telford in October 1811 and valued it. The price they offered for the land was rejected. An encounter with a landowner like Glengarry must have been a unique experience for the Canal-builders. Alastair (or Alexander) Ronaldson MacDonell of Glengarry lived like a seventeenth-century chief unaware that he had strayed into the nineteenth century. Arrogant, impetuous, obstructionist, he always travelled accompanied by his 'tail' of clan supporters and expected to get his own way in everything. Short of money, he was quick to realise that the needs of the Canal works immediately gave his forests a new value and had provided timber worth £1,791/18/- by the spring of 1805. If the Canal Commissioners had been able to

Opposite: Alastair MacDonell of Glengarry by Sir Henry Raeburn.

anticipate his intransigence they might have been slower in settling this account. He demanded that the amount of compensation due to him for his land be assessed by a jury. In 1813 George Brown walked the lands again, accompanied by Matthew Davidson and William Hughes, the contractor, and prepared a new valuation. The following September, Brown, Telford, James Hope the legal agent, and Henry Cockburn, their advocate, were in the Highlands to agree how best to present their case to the jury. Later when he was Lord Cockburn and a criminal judge recording his travels round the country in his *Circuit Journeys*, Cockburn recalled his first encounter with Glengarry:

> The only fearlessness he ever displayed was in an act of madness which Telford (the engineer) and I saw, and to which he was driven by insolent fury. A boat in which he wished to cross Loch Oich, or Loch Lochy, left the shore without him, and a laugh showed that it had done so on purpose to avoid him, on which he plunged with the pony he was on, into the water, to swim after it. The people pretended not to see, and rowed as hard as they decently could. Telford and others were in ecstasy with the hope that they were at last to be relieved of him. And certainly he ought to have drowned. But after being carried very nearly across (a mile I should suppose), by the vigour of a creature more meritorious than its rider, he got on board, and was praised for what it had done.

Up near Loch Oich, Brown and the jury were soaked as they trudged along the line of the Canal in wind and rain. All the next day until late in the evening in the Fort at Fort Augustus the jury heard witnesses, many of them Gaelic-speakers from the west coast. They reached their verdict at four in the morning and awarded Glengarry almost £10,000, which was '£3,000 more than he was entitled to, had they been honest men', as George Brown noted disappointedly in his *Journal*.

The award failed to satisfy Glengarry. In 1816 he and thirty armed men drove off Canal workmen and seized one of their boats. He demanded that ships keep to the south shore of the Loch, away from his house, and that a bank be built to screen his house from the Canal. His claims went on for years, about the siting of accommodation bridges, compensation for bridges not built in places which would have suited him and for damages for the right to navigate Loch Oich. As late as 1826 he estimated his claim for damages to Loch

Oich at £20,000 in addition to proceedings he had already instituted against the Commissioners in the courts. In truth, he resented the construction of the Canal and the prospect of 'passenger-boats and smoking steam-vessels' which he regarded as a threat to a way of life which, had he but realised it, had already passed away. That is why trouble was expected from him on the ceremonial voyage in 1822 and the voyagers suddenly changed their plans and left Fort Augustus at such an early hour as six in the morning of 24 October in case he tried to prevent their entry to Loch Oich by armed force. They need not have troubled themselves. He joined them on the way and at Fort William attended 'the handsome and plentiful dinner', where he proposed three toasts and responded to two before retiring early. A heroic figure in some ways, who had raised the Glengarry Fencibles during the war, he was greatly admired by Sir Walter Scott who described him as 'a much misunderstood man'. Telford and the Canal Commissioners had reasons in plenty for finding him difficult to understand.

Except for the lock-building at Fort Augustus, all the final Canal work was in Glengarry's lands, about two miles of cutting at each end of Loch Oich, as well as the deepening of the Loch. On the Canal line at the east end of Loch Oich, William Hughes began work in 1815 on a small scale because funds were low. He had to build up the west bank of the Canal first of all to protect it against the River Oich and then to remove an island and cut a new channel for the River to prevent it flooding the Canal and the land around. Big stones from the cutting were used to construct the river bank. Not till 1821 was a suitable solid rock foundation found on which to construct the last of the locks, Cullochy regulating lock. Like Kytra Lock it was built without an inverted arch. Durable granite found nearby proved difficult to quarry and difficult to work but it was persisted with, since the alternative was the delay and expense of bringing in better-quality stone in quantity from Redcastle, over forty miles away. It was tedious enough to have to import lime and some freestone for the hollow quoins. The lock was not completely finished until after the opening of the Canal, when it was easier to bring in the final cargoes of freestone by water. What it was like to work in such an out-of-the-way place was recalled by Joseph Mitchell, son of John, Telford's principal inspector of roads, in *Reminiscences of My Life in the Highlands*:

I lodged with about thirty masons in a house built for the lock-keeper [at Cullochy.] The men slept in temporary beds, one above the other, like the berths on a ship. The men began work at 6 a.m. one being told off to cook about half an hour before meals. The breakfast was at nine o'clock, the fare consisting of porridge and milk and thick oaten bannocks. They dined at two on the same fare, and at eight had supper. The fare varied when the new potatoes came in and fresh herrings were brought in from Loch Hourn in the autumn. On Sundays they luxuriated in tea, oaten bannocks and butter for breakfast.

Sunday was reverently kept. The men were perfectly sober, never tasting spirits or beer, and as the cost of living amounted only to 3/6d or 4/- a week out of wages of 21/-, each had a very considerable sum to bring home to his wife and family.

The Highlanders who were labourers lived chiefly on brose, i.e. meal in a bowl, a little salt, and hot water mixed into a mess. There was little or no drinking among them here for there was no public house within three miles of the place; but at Fort Augustus after a pay-day, which was once a month, they took to drinking and quarrelling, and spent then as much money as would have fed them comfortably the whole month.

Middle West – Loch Lochy to Loch Oich

The transfer of workmen from the west to the east end of Loch Lochy was like an invasion of the highest land the Canal was to pass through. The Loch Lochy barge did journey after journey in the spring of 1814 carrying forward tools and timber. The contractor, John Wilson, built timber-framed houses with slated roofs, as well as workshops for the carpenters and smiths, while the labourers 'hutted themselves as usual' in stone or turf houses. Wilson had the let of the Canal land and used it for grazing his horses and cows and producing potatoes for the men and fodder for the animals.

About a hundred and fifty men started on the deep summit cutting at Laggan between Loch Lochy and Loch Oich. The work went on for years. After they had done half the digging that was needed they were down only to what was to be the surface level of the Canal. The deeper they excavated, the more difficult it became because of the height of the Canal banks. In the end, the bottom was made only thirty feet wide instead of the usual fifty. Near Loch Oich excavation had to

The tree-clad banks of the Canal at Laggan Summit.

stop at the summer level of the Loch until Laggan Locks were built and they depended on the arrival on the steam engine from Corpach before a start could be made. Much of the earth was moved by horses and waggons, along the railways. At the lowest depth the soil was to be excavated later by dredging machine when enough water could be let into the Canal to float it.

The locks at Laggan were built on a foundation of decomposed brushwood and vegetable matter which had become so hard that it could not be penetrated by water and the workmen could hardly cut it with their picks. The massive scale of the works amazed Robert Southey in 1819 as he looked down from the top of a mound of excavated soil into the lock-pit where the men seemed to him to be 'in the proportion of emmets to an ant-hill amid their own work'.

Looking down on Laggan Locks.

Most of the stone for lock-building came along a railway from the Kilfinnan Burn, fairly near at hand, but good freestone for quoins and cope-stones still had to be brought all the way from the Firth of Clyde. In 1821 when Laggan Locks were ready and the dam in the old River channel was built up to its full height at the south-west end of Loch Lochy, the level of the Loch was raised by ten to twelve feet and the River Lochy transferred to its new bed. The Loch rose beyond that level because the new watercourse for the River under Mucomer Bridge was too narrow but the dam stood the strain. The rock was cut wider and deeper to give the River more room to tumble freely into the Spean. The west end of the Canal was now open from Corpach to Loch Oich.

Loch Oich

The shallow parts of Loch Oich were a much more difficult dredging problem than those in Loch Dochfour had been. Jessop had designed the machinery for a steam dredger for Loch Oich as early as 1805 but the barge sank and was left there. Later it was decided to lower the

A sailing ship being towed along Loch Oich, Invergarry Castle on the right.

level of the Loch by draining water off to allow some cutting to be done by pick and shovel. It looked as if the problem of dredging this inland Highland loch was almost beyond him but his contacts with other engineers in the south proved vital in this case. The number of civil engineers in Britain was still small: they knew one another and knew who would be most likely to find solutions to their problems. This explains why Bryan Donkin, a native of Northumberland and a founder-member of the Institution of Civil Engineers, was up at Loch Oich with Telford in 1816. 'The ingenious Mr Donkin of Bermondsey', as Telford called him, was capable of applying an alert mind to problems in almost any field. His inventions practically started the machine-made paper industry and another of his ideas, an airtight case for preserving meat and vegetables, was proving a boon for ships' crews on long voyages. Telford and Donkin realised that they needed a more powerful dredger on Loch Oich than had served on Loch Dochfour. Donkin designed the dredging machinery which was made in the south while Thomas Rhodes' carpenters practically introduced shipbuilding to Loch Oich when they constructed the 216-ton dredging vessel *Glengarry* of oak timber, along with four 50-ton

The plan, discovered only in 1992, which George Brown made of Fort Augustus in 1811, showing the Fort and the land belonging to it, the old course of the River Oich and the new line proposed for it, and the intention to provide a passing place between the lower two locks and the upper three.

discharging barges. Equipped with two sets of strong buckets, powered by a 10 hp engine, the dredger was in the charge of Alexander Fyfe, who later became the first engineer on the Liverpool and Manchester Railway. The *Glengarry* proved capable of extracting 800 tons in twenty-four hours. It even pulled up great oak trees from the bed of the Loch.

Robert Southey, the Poet Laureate, who accompanied Telford on his inspection in 1819, saw it in action: 'We went on board and saw the works; but I did not remain long below in a place where the temperature was higher than that of a hot-house, and where machinery was moving up and down with tremendous force, some of it in boiling water.' Later a local young man was killed by its boiler exploding.

Technological invention in the form of the steam engine for draining lock-pits and driving dredging machinery allowed Telford to construct public works which would have been impossible earlier. On the other hand, steam-power for propelling ships was beginning to outdate some of the ideas on which the Caledonian Canal was based. For example, all along the south-east side of the Canal a towpath had been carefully constructed for the use of trackers with horses to pull vessels against contrary winds. Would steamboats as tugs not be preferable? Telford asked Donkin to consider how the Canal's stock of steam-engines might to used to power steamboats and decided to choose a route through the deeper centre of Loch Oich and dispense with a towpath along the southern shore. Glengarry could not have welcomed his decision.

Fort Augustus

Down at Fort Augustus, the course of the River Oich where it entered Loch Ness was moved completely to the north-west side of the little island there and the whole bank was dammed and secured. The line of the locks having been fixed through land belonging to the Fort, the great 36 hp Boulton & Watt engine, which had been stored unused at Clachnaharry for ten years, was brought up Loch Ness and installed at Fort Augustus to pump out water while the lowest lock was being excavated twenty-four feet below the level of the River and the Loch. Water penetrated the gravel bank so easily that sinking an engine pit to a depth of twenty-eight feet proved almost impossible.

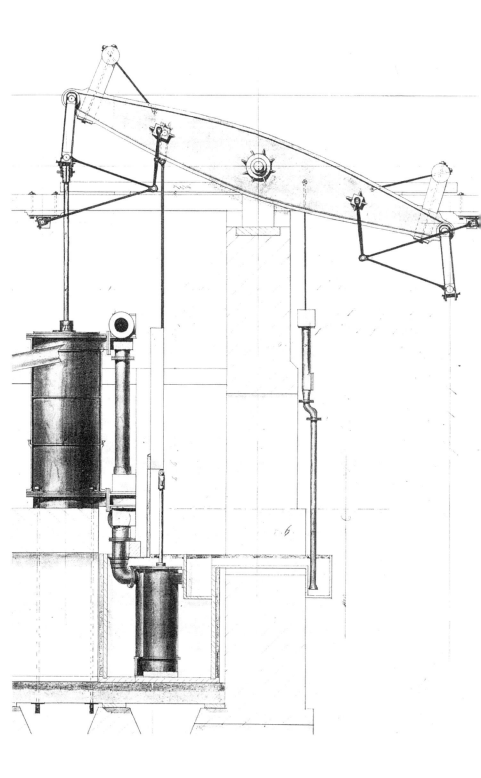

Telford declared his intention of bringing up other steam-engines from Clachnaharry and Corpach if necessary.

Fortunately the engine was powerful enough to allow excavation to proceed. Making 15 to 17 strokes per minute, it discharged over eight hogsheads of water at every stroke (a hogshead is 52½ gallons) and consumed 33 barrels of coal every 24 hours. By autumn 1816 labourers were hard at work twenty feet below the surface level of Loch Ness. When it came to building the foundations of the lock, however, three steam-engines with a combined force of 62 hp had to be employed in the greatest pumping operation in the construction of the Canal. By comparison, the Corpach sea-lock had been kept dry by a 20 hp engine and the one at Clachnaharry by a 6 hp engine. When bottom-level was reached, the masons moved in and with all the engines working to keep the masonry as dry as possible, they laid the first course of flagstones packed in moss, and the second in mortar made of three parts sand to one of lime. Then they built the inverted arch and platform but the lower sill was set too high because they could not get down for water. The lock could only give a rise of 7 ft 4 in., instead of 8 ft, but the difference was divided among the remaining locks above. In spite of this Telford, on his inspection in October 1817, pronounced his pleasure at the progress made:

> By great exertions, and two months of favourable weather, this has been accomplished in a very perfect and workmanlike manner . . . The most arduous and uncertain portion of this work at Fort Augustus has been accomplished.

When the guns and the old soldiers departed from the Fort in 1818, the Commissioners tried to take it over as warehouses but determined opposition by the Duke of Wellington defeated them. The Canal-builders became the new centre of interest in the village. There were about two hundred of them that season: masons and labourers in the lock-pits, carpenters and blacksmiths in the workshops beside the bottom lock, and masons and quarrymen in the quarry on the north side of the River. Some horsemen brought rubble-stone in their waggons along the tramway from the quarry, others collected

Opposite: Side view of one of the Boulton & Watt steam engines which would have been pumping here, the drawing 'to be kept clean, not permitted to be copied, and returned as soon as the Engine is finished.'

freestone and lime from vessels which discharged their cargoes at the Fort pier. Freestone came from Ruskich quarry on the north side of the Loch and lime from the same area. Now, the masons worked in the winter months as well, getting stone ready in the quarries, even across at Redcastle. When the Canal was opened from Clachnaharry to Lock Ness in 1818 sloops from Redcastle could then bring freestone all the way by water to build the inverted arches, the hollow quoins and platforms, for which the Ruskich stone was not suitable.

Robert Southey described the scene with boyish delight:

Men, horses and machines at work; digging, walling and puddling going on, men wheeling barrows, horses drawing stones along the railways. The great steam engine was at rest having done its work . . . The dredging machine was in action, revolving round and round, and bringing up at every turn matter which had never been brought to the air and light. Its chimney poured forth volumes of black smoke, which there was no annoyance in beholding, because there was room enough for it in this wide clear atmosphere. The iron for a pair of lock-gates was lying on the ground having just arrived from Derbyshire:

Sailing along the Laggan avenue.

the same vessel in which it was shipped at Gainsborough, landed it here as Fort Augustus.

The construction of the five locks went on steadily uphill but the idea of creating a pound between the lower two locks and the upper three, which can be seen in George Brown's plan dated 1811 (p. 84), was abandoned. The masons were furthest ahead with the stonework in the lower locks, while the labourers were excavating the locks above them and taking the soil down to build up the banks behind the completed lock-walls. By the spring of 1821 the masonry on the five locks was complete, the lock-gates were in position and the steam-engines, so essential to the early stages of the works, were silent. One of the engine-houses still stands. Cargill's masons had built 300 yards of lock masonry, and in that distance the Canal could raise ships forty feet above the level of Loch Ness.

The village benefited from the spending power of the workers and began to share in trade with the world outside. For the first time the timber of the Great Glen could be exported and in 1820 one and a half million birch staves were sent out for making herring barrels. Incoming coals became cheaper and were burned in some Fort Augustus fireplaces for the first time. A steamboat, the property of Henry Bell who had pioneered the first steamboat passenger service between Glasgow and Greenock on the Clyde in 1812, plied regularly between Inverness and Fort Augustus. It was possible for a traveller to sail from Inverness to the Fort, travel by stagecoach to Fort William and sail from Fort William by way of the Crinan Canal to Glasgow. Even in its unfinished state the Canal put Fort Augustus on the traveller's route and brought Inverness into closer contact with the western Lowlands.

9

COMPLETION UNDER CRITICISM

When the Napoleonic War ended in 1815, one of the reasons for building the Canal – to facilitate the movement of 32-gun frigates from sea to sea – ceased to have any validity. The strain of a long and arduous war, which had raised the National Debt to the unparalleled height of £861 million led to demands for retrenchment and the reduction of taxation. Some writers, such as James Robertson in his *General View of the Agriculture of the County of Inverness*, had marvelled at the liberality of Government expending £50,000 a year on this Highland canal while conducting a major war in Europe. Now there were serious questionings. Why had the Canal cost so much more than the original estimate already and how much more was it going to cost? Why was it still not ready and how many more years was it going to take? Doubtful of its commercial viability, other critics suggested that Government grants should cease and the works be abandoned.

In this climate of mounting public criticism, the Canal Commissioners were driven to make forecasts which were too optimistic and suffered further onslaughts when their predictions were not fulfilled. In 1816, for example, they considered that the Canal would be completed in three more years: in fact it took six. In the same year, arguing that the Canal's annual income from tolls would probably be in excess of £40,000 a year, they asked for the grant for the next three years to be paid in two annual instalments of £75,000 in order to bring the Canal into operation sooner. To ask for a higher grant was wise because in the later stages so much more of the money was needed to pay for timber and machinery that only about half was left to pay wages. The forecast of future income was wildly optimistic, since annual tolls did not even produce £3,000 until the late 1840s.

Telford's estimate of the amount required in 1816 to complete the Canal was £171,000. After receiving grants of £150,000 by 1818, he reckoned that a further £80,000 was necessary. Called on to account for the continued increase in costs, he drew attention to the rises in the prices of food and materials and in wages of the order of 30 per cent to 50 per cent. In 1804 labourers had been paid 1/6d, and in 1812 2/4d to 2/6d for day work, while the piecework rates paid to labourers per cubic yard had gone up from 3d to 4½d. The same carpenters, who had received between 2/3d and 2/6d a day in 1804, were paid from 2/10d to 3/4d in 1813. The same masons had seen their weekly wages rise from 16/- to 21/- by 1813 while blacksmiths were earning a shilling a day more. As an example of the rise in the cost of food, Telford pointed out that oatmeal which cost 2/6d to 2/7½d a stone in 1803 had risen to 3/3d in 1811, 4/6d in 1812 and 4/9d in 1813. Materials such as timber were much more expensive. Fir purchased locally at 10d to 1/2d per cubic foot in 1804 cost 3/6d per cubic foot ten years later. Baltic timber landed at Corpach had risen in price from 2/6½d to 7/- per cubic foot while the price of oak had almost doubled to 12/- and was difficult to obtain. The substitution of cast-iron for oak for making lock-gates practically doubled their capital cost but he reckoned they would last much longer. The examples he gives of the rise in the prices for food and materials suggest that his figure of 30 to 50 per cent was an under-estimate of the extent of the increase, while the greater percentage increase in wages received by labourers than by craftsmen must have been the result of the higher costs of basic foodstuffs.

The amount paid in compensation for land was much higher, about three times higher, than had been expected. Clay lining of the Canal, to prevent leakage from certain gravelly stretches on to adjoining property, had proved time-consuming and very expensive. In the south-west, far more rock had had to be cut through than the trial pits had given reason to believe. Perhaps the principal cost which he had under-estimated was for inventing and constructing dredging machinery and operating and repairing it, especially at Loch Oich. On the other hand, he considered that the capital equipment, such as steam engines, dredgers and cranes, and the land which the Commissioners now owned, were worth at least £50,000.

In addition to these explanations given by Telford, the assembly and operation of three steam-engines at Fort Augustus cost far more

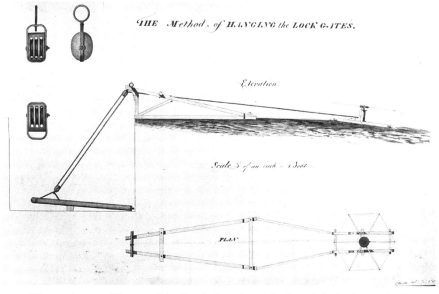

The equipment Thomas Rhodes designed and built to lift the heavy lock-gates into position.

than anyone could have foreseen. The defensive tactics of the Board of Ordnance who objected to the Canal works passing through their land at Fort Augustus, and the difficulties in finding a suitable site for the locks there, upset the phasing of work and added to costs. When most of the work was going on in the Middle District, supervision from bases at either end cannot have been so effective. Probably too the distance from the ironworks in Derbyshire and Denbighshire, which supplied cast-iron for lock-gates for example, caused problems of communication between customer and supplier, as well as making delivery slow and unpredictable. As the years passed, the main contractors on the Canal were undertaking other commitments over a wide area. William Hughes, for example, was involved in searching for coal at Brora in Sutherland. Simpson and Cargill built Bonar Bridge over the Kyle of Sutherland fifty miles north of Inverness and Craigellachie Bridge over the Spey forty miles to the east. While these undertakings may be interpreted as signs of the expansion of these early firms of contractors, they meant that key men were not always available on the Canal, and that the Canal was no longer the absolute priority it had been to them in the beginning. The early impetus was fading. There is no doubt that Telford missed the expert advice of Jessop after 1812, and no one was appointed to replace him. Telford, who in the early years had been up at the Canal two times every year, now made only one visit a year, lasting about six weeks, and fitted in among his own ever-growing commitments. In that time after 1819, without Jessop, Simpson and Matthew Davidson, he had to inspect the works and accounts and take vital decisions. Of course, other men, such as Thomas Rhodes, were growing into responsibility. Rhodes, who had studied mathematics and drawing under the rector of Inverness Royal Academy, made drawings of the solid lock-gates and worked out and built the equipment to hang them; but James Davidson, the superintendent of most of the current work, was young and, as time was to prove, not the fittest of men.

Suggestions were beginning to be made that it was not Highlanders who were building the Canal but Irishmen. Lord Carhampton put this view most sweepingly in the House of Commons when he maintained that 'all the benefit had been reaped by Ireland'. Many years later, in 1841, the *Select* Committee on Emigration from the United Kingdom argued from the presence of Irish squatters at Corpach and in huts along the line of the Canal that the Irish took over and that local labour ceased to be employed. The official reply, prompted by the accusations in Parliament, was published in the 1819 Report. For Clachnaharry, James Davidson provided the information that the number of English workers was never more than three, Welsh four and in later years only one, and Irish only one, except for eight who worked for four months in 1813. Alexander Easton reported that at Corpach there were three Irish labourers, three English labourers and two English carpenters, the Rhodes brothers, and that all the rest were Scots. When work started, no Irishmen had appeared for work but in 1811 and 1812, about one worker out of thirty was Irish, or thirty-four in the peak month, August 1811. The labourers, according to Easton, were almost entirely Highland from Inverness-shire and Argyll, mainly from the north side of the Canal, with others coming from the Islands: Skye, Lewis, Mull and Lismore. His masons were from Moray and the blacksmiths and carpenters from around Inverness. There was an influx which probably included some Irishmen from the Glasgow area in the depression of 1811 and 1812 but, even if the superintendents' figures err on the conservative side, there are no grounds for suggesting an Irish takeover of the Canal work. The evidence points the other way. Telford's account of the rise in wages up to 1813 mentions specifically that the carpenters and masons in 1813 are the same men who were at work there ten years earlier. In a letter in 1818 Telford deplored the departure of Highland labourers on their seasonal pursuits:

> The herring season had been most abundant, and the return of the fine weather will enable the indolent Highland creatures to get their plentiful crops and have a glorious spell at the whisky-making.

This outburst, and Matthew Davidson's acquired antipathy to Highlanders would have appeared superfluous if the bulk of the workers had been Irish. Nor did they ever criticise the Highlanders adversely by comparing them with superior workers such as the Irish. Lastly, the list of lock-keepers in 1826, one of the few lists we have of men who had worked on the Canal, contains few, if any, Irish names. All this proves is that the Canal was built mainly by Scottish workmen.

Opposition in Parliament was parried and grants continued till 1820 when £60,000 was made available but nothing was forthcoming the next year. For the next three, £25,000 a year was granted, even although in 1822 the Canal was officially open.

Almost from the beginning of the Canal project, Jessop's and Telford's ambition was to bring each part of the work to completion. In the annual reports, the favourite word seems to have been 'complete', applied to masonry and to stretches of cutting. To their minds the Canal was not 'complete' when it was deep enough and the banks the correct height. The banks had to be bound together to keep them secure. One of their instructions to Matthew Davidson in 1810 was 'to collect broom and whin seeds and sow them on the slopes of the Canal, especially on the inside'. Broom and whin, direct descendants of that seed purposely sown, flourish on stretches of the Canal today. Grass was sown too on the banks, and above Fort Augustus the banks were turfed. By the opening date £108/18/8d had been spent on grass, broom and whin seed. With recent memories of the trees planted to hold the soil on William Davies's great mound

Opposite: Gatelifter III taking on a gate leaf at Muirtown.

at Pontcysyllte in North Wales, Telford ordered trees to be planted on a massive scale along the Caledonian Canal. Some of them were Norwegian spruces, grown from seed presented to him by Count Von Platen, whom he had advised about the construction of the Göta Canal in Sweden. Twenty thousand fine thorns, which cost £10 from Dickson & Gibbs, Seedsmen in Inverness, were planted at the west end in 1819. In all, some half a million trees were planted and were flourishing, serving three purposes – binding the banks, growing into valuable timber, and landscaping the Canal. Such a massive public work would otherwise have been a great scar along the Glen but he was creating a new landscape to blend with the old. The cost was small, £485/12/11½d, not counting labour, and the 1824 Report commented that 'the Plantations, which are extensive, thrive remarkably well'.

Aware of criticism, but not overwhelmed by it, Telford was influenced to accept lower standards with regard to the breadth and depth of the Canal in the Middle District. The intended depth of twenty feet was not achieved by then but dredging, he felt, would cure the deficiency in time. He had also to allow poorer building stone to be used because it was available nearby, in the interests of getting the Canal opened soon. Back in 1807, the Clachnaharry lock, now the Works Lock, appealed to him as the finest example of the lock-builder's art: in the later years in the centre he was impressed – and perhaps had to be – by speed of execution. Considering that for a vessel passing through, the depth of the Canal is the depth of the shallowest part, and the safety of the locks the soundness of the poorest stonework, the Caledonian Canal did not measure up to the felicitous phrases used in the Annual Reports.

The pressure to have the Canal completed kept the number employed at a high level reaching a peak since 1812 of nine hundred and forty-six in July 1819, when there were nearly two hundred men at Loch Oich, two hundred and fifty at Fort Augustus, and two hundred and seventy-five at Laggan. Numbers fell in the winter to two hundred and fifty to three hundred when no masonwork was done but in the next two summers the total number was back up to about seven hundred. In the long days of 1822, just before the Canal opened, over four hundred and fifty men were still at work, mostly between Fort Augustus and Loch Oich. Then in spring of 1823 the figure fell to a Canal establishment of sixty-nine, but not for long. Essential minor repairs and maintenance meant employment for over two hundred men that summer, and over three hundred the next.

In May 1823 it was time to count costs: £912,373/8/7½d in all had been received by the Canal Commissioners and all of it except about £7,000 had been spent on the Canal. In relation to the total expense, the cost of management at £29,000 was not high. The amount received by Thomas Telford for almost twenty years of inspection, direction and worry amounted to less than £5,000, and Jessop, for ten years, £2,057. The total cost of the salaries of the resident engineers at Clachnaharry and Corpach was just over £9,000. John Rickman, the Secretary to the Canal Commissioners, conducted all their business for a salary of £200 a year and all their legal work in Scotland was done by Mr James Hope at a cost of £1,870 for the whole period. George Brown earned £423 for drawing plans, making valuations and attending juries. Among the smaller items, Murdoch Downie's survey of the lochs in 1803 cost £110/11/3d, John Howell's survey and maps in 1803–4 £186, and A. Arrowsmith's drawing of bold maps which showed the progress of the work in the Annual Reports by the use of colours, earned his fee of £78/2/-. Captain Mark Gwynn

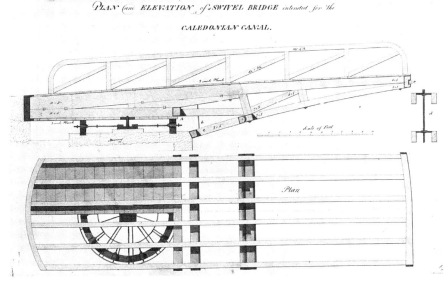

Thomas Rhodes's drawing of one half of a swivel bridge across the Canal.

kept the record of winds and weather at Fort Augustus for six years for £35/14/-.

The biggest item of expenditure was naturally the payment of labour, amounting to £418,000, divided roughly equally between the three districts for excavating, puddling, banking and laying railways and for the work of sawyers and carpenters, but not masons. The cost of quarrying and masonwork amounted to almost £200,000. These two items taken together show that two-thirds of all the money made available by the Treasury found its way into the Highland economy in the form of wages.

Compensation to landowners accounted for £47,887, including £10,477 to Glengarry. Some estate improvement resulted from this infusion of money while some landowners used it to pay their share of the cost of the roads being built through their lands. Glengarry, in fact, was relying on it for his contribution to the Glengarry road. The other fairly large expenditure in the Highlands was for local timber costing £27,764 compared with the sum of £40,813 spent on English and foreign timber. It is interesting that William Hazledine, the ironmaster of Plas Kynaston, who supplied the ironwork for the lock-gates and bridges at the west end, actually received more for the oak he provided than for iron – £15,000 compared with £11,000.

Machinery, cast-iron and equipment cost over £121,000. Most of it was purchased outside the Highlands but it did include £9,396 for iron rails and other ironwork from Jonathan Wells's Inverness New Foundry. This was an outstanding example of local enterprise being sparked off by public works and being able to compete in terms of quality and price, aided by much lower delivery costs. Outram's Butterley Works in Derbyshire had provided iron rails and wheels worth £4,687 by May 1806 and orders in the next seventeen years, including dredging machines, cost a further £4,000. The Inverness New Foundry first appeared in the accounts in 1809, providing iron rails etc. for the sum of £886/7/9¼d. Then it became the main supplier of rails and castings for the lock-gates and by 1820 its income from Canal work amounted to about £100 a month.

Butterley also supplied all the cast-iron bridges and cast-iron framing in place of oak posts and ribs for the locks at the east end and in the centre at a cost of £18,023 while Hazledine did similar work in North Wales for the west end worth £11,253. Carriage of ironwork and machinery to the Great Glen added £6,500 to costs. For example, William Young received £121 for transporting two cast-iron bridges and their timbers from Bristol to Corpach in 1820.

The first and most significant order for machinery was for three steam-engines from Boulton & Watt in Birmingham, which may be said to have introduced the steam-engine to the Highlands. The engines, 36 hp, 20 hp and 6 hp, and spare boilers for them were delivered at a cost of £5,596. The accounts show that the engines proved more expensive to operate than to buy, with sums of over three thousand pounds for building engine houses and over six thousand pounds for coals.

The two sloops for carrying stone from Redcastle, one called the *Caledonia* of Inverness, were built by Samuel Deadman and John Nichol at Inverness while for the west end, the *Caledonia* of Fort William was the work of John Stevenson of Oban and the *Corpach* of William Courtney of Chester. The sloops on Loch Ness and Loch Lochy and the dredging vessels on Loch Oich were built by the Canal carpenters themselves.

The expense incurred in housing men and horses and constructing the other buildings needed for the progress of the works is calculated in terms of labour costs only and amounted to just over £4,500:

Houses for Workmen and other Temporary Buildings:– Workmanship	£	s	d
House for Superintendent at Clachnaharry, including House for Superintendent's Clerk	657	13	9
House for Contractors, Overseers, and Workmen, and Timber Yard in the Clachnaharry District	835	10	3
Stable at Clachnaharry	170	3	8
Fitting up house for Overseer at Fort Augustus (1818)	71	19	10
Sheds for workmen in the Middle District	693	1	8
House for Superintendent at Corpach	360	8	7½
Houses for Contractors, Overseers and Workmen in the Corpach District	1,124	3	0½
Stable, Granary, and Office at Corpach	318	17	0½
Brewhouse at Corpach	130	5	4½
Stable at Muirshearlich	44	7	11
Stable at Mucomer	178	4	6¼
Total Expense of Workmanship on Houses for Workmen etc.	£4,584	15	9

Four thousand pounds was spent on roadworks. Three thousand pounds only appear as the cost of horses and fodder because

Moy Bridge, the only original cast-iron swivel bridge on the Canal now, is still opened one half at a time, by hand.

contractors were usually expected to buy and feed their own work-horses. £127/7/0½d was expended on whisky, well illustrated by one account in November 1811 for '£1/6/8d for 5 pints of whisky given Men when out at the water ditch and Engine at [Clachnaharry] Sea Lock at nights'.

Criticised by many on grounds of cost and referred to by some as a 'Scotch job' to keep Scottish MPs loyal to the Government, the project was defended in an article in *Blackwood's Magazine* in July 1820, because of the great change it had achieved in the attitude of the Highlanders to work, the skills they had acquired and the capital they had accumulated, enabling some to take up handicrafts, others to buy fishing boats. The suggestion that the Canal was too big for its location was refuted, by reference to the ditch canals of England and the Forth and Clyde Canal, which were already too small. The Canal was described as a great national enterprise undertaken for the prosperity of the whole of the United Kingdom. Nonetheless many doubted whether such a big canal would ever work and the delays over completion made them feel that they were right. The opening in October 1822 at last proved them wrong.

10

WORKING THE CANAL – IN THE 1820s

Once the Canal was open for traffic it needed at least a team of men to operate the locks, a table of rates and a set of rules and regulations. The Caledonian Canal Act allowed the Commissioners to charge up to twopence per mile per ton of cargo carried along the Canal and also to charge unladen ships up to the amount of cargo they were capable of carrying. As an incentive to encourage the owners of larger vessels to use the Canal, the Commissioners had the power to permit vessels of 100 tons burden to pass through free of charge provided they went from sea to sea in the first three years the Canal was open. The evidence available to them suggested that the Pentland Firth was a treacherous channel where strong winds made ships run for shelter and where many ships were lost. In a letter to Telford an Aberdeen shipowner, John Saunders, mentioned specific losses such as two cargoes of coffee and sugar being carried from Liverpool to the Baltic in 1813, one of them valued at £100,000. Ships from Aberdeen, he said, could be delayed sheltering in Orkney for as long as it took them to cross the Atlantic and he thought that by using the Caledonian Canal, Aberdeen ships could make two voyages a year to America instead of only one.

The Commissioners who were already charging 6d per ton for the voyage to Fort Augustus from Inverness decided that no ships should pass through free. It seemed better to have uniform rates and to keep them low:

> On all Vessels navigating the whole or any part of the Inland Navigation . . . shall be charged and payable a Tonnage Rate of ONE FARTHING PER TON PER MILE, calculated upon the Registered Tonnage . . ., and in case of any voyages so short that the foregoing rates of Tonnage shall not amount to SIX PENCE PER TON, so much is to be charged . . . at TWO PENCE PER TON PER MILE, as shall amount

to THREE PENCE PER TON INWARDS, and THREE PENCE PER TON
OUTWARDS.

The Sum of FIVE SHILLINGS to be paid at the Muirtown Locks
for every voyage upwards of any Steam-Boat, and the same sum for
every voyage downwards; and FIVE SHILLINGS upon passing the Fort
Augustus Locks; no charge whatever . . . on Steam-Boats passing
inwards or outwards at Corpach. Thus a Steam-Packet will be charged
Ten Shillings for the entire passage from Sea to Sea.

On all Vessels taking in or discharging their Lading at the Wharf
of the Muirtown Basin near Inverness, and the Corpach Basin near
Fort William, shall be charged and payable a Tonnage Rate of One
Penny per Ton.

For the sake equally of the Lock Gates and Bridges, and of the
Vessels which navigate the Canal, all such Vessels shall take on board
an authorised Pilot, if required so to do by the Collector of Tolls.

There is no want of Horses for tracking Vessels along the Canal,
when the wind fails, or is adverse. The rate of going is usually Two
Miles per hour; the payment for the Horse Five Shillings per day; the
distance between Corpach and Loch Lochy being reckoned a day's
work, and between Clachnaharry and Loch Ness the same. Steamboats will be placed in the Lakes as soon as there is a demand for their
services in towing heavy Vessels.

If tolls were to be based on mileage as well as tonnage, milestones had
to be erected along the line of the cutting. Then the poetic tribute by
Robert Southey to Telford and his works was to be inscribed on three
stones and mounted at Clachnaharry, Fort Augustus and Banavie as
a surprise for Telford. It included these lines:

Telford it was by whose presiding mind
The whole great work was planned and perfected; . . .
Telford who o'er the vale of Cambrian Dee
Aloft in air at giddy height upborne
Carried his navigable road; and hung
High o'er Menai's Strait the bending bridge:
Nor hath he for his native land performed
Less in this proud design: and where his piers
Around her coast from many a fisher's creek
Unsheltered else, and many an ample port
Repel the assailing storm: and where his roads
In beautiful and sinuous line far seen

Wind with the vale and win the long ascent
Now o'er the deep morass sustained, and now
Across ravine, or glen or estuary
Opening a passage through the wilds subdued.

James Davidson wrote to John Rickman, 'There are no tradesmen here who can engrave the Inscription, as it ought to be. I think one large white Marble slab can be had from Leith Walk, Edinburgh'. It was purchased and engraved but not erected due, it was said, to the engineer's modesty. Telford was aware that the planning of 'the whole great work' owed more to Jessop than to himself in the early years and he knew too that the high aqueduct 'o'er the vale of Cambrian Dee' had also been their joint endeavour. This stone, meant for Banavie, was discovered and mounted on the wall of the old Canal Office at Clachnaharry in 1922, where it is still on view.

Thomas Telford advised that steady, experienced men be appointed as lock-keepers and that care be taken in the management of the flights of locks. Pilots were to be appointed at Corpach and Clachnaharry, preferably men who had previously been masters of Canal sloops. James Davidson, superintendent at the east end became Resident Engineer for the whole Canal, while his opposite number, Alexander Easton departed for a canal appointment in Ireland. Davidson had been a pupil of Telford's and was, according to Telford, 'not the worse for the 12 months discipline I afforded him'. At Corpach, Alexander Clark was Collector of Dues and Andrew May, whose son later became Resident Engineer, held a similar position at Clachnaharry. Each was paid £150 a year, half the Resident Engineer's salary. The Superintendent of the Middle District, Alexander Fraser, received £78 a year, the same salary as James Rhodes, the carpenter who became principal keeper at Banavie, and Thomas Jones, carpenter and lock-keeper at Clachnaharry. In every case, the man in charge of a lock or flight of locks was a carpenter, e.g. William Fraser at Corpach, James Rhodes at Banavie, Simon Fraser at Fort Augustus. His assistants had usually some different skill, for instance, at Corpach a blacksmith and a labourer, at Banavie, a mason and a shipwright as well as John Macphie, seaman and pilot, and some labourers. An interesting assistant who was still at Fort Augustus in 1826 was James Fyfe, keeper of steam-engines, whose brother had been in charge of dredging machinery on Loch Oich. Fyfe was paid £72/16/- a year, almost as

much as his superintendent, and two pounds a month more than the principal lock-keeper, but there was now no work for him to justify his old rate of pay. John MacLucas, an assistant at Clachnaharry, became seaman and pilot at 15/- a week. Masons generally received 15/- to 18/- a week, blacksmiths 15/- and labourers 14/-. One or two labourers who were not permanently on the Canal establishment were paid 10/- or 11/- a week. On the whole these wages were lower than those paid to workmen during the period of construction but the new rates were paid all the year round and usually the lock-keepers were provided with houses. Some had other means of augmenting their income. The lock-keeper at Dochgarroch farmed twenty-four and a half acres for a rent of £15 a year and the Inspector at Fort Augustus and the Collector at Corpach both had farms of fifteen acres each. James Rhodes, the principal lock-keeper at Banavie, was allowed to run one of the lock-keepers' houses as an inn for the convenience of steamboats' passengers. This arrangement and his appointment ended in 1834 because of his alleged abusive conduct towards masters of vessels who did not frequent the inn. There were even suggestions that some skippers had sailed through the Pentland Firth to avoid it!

The allocation of workers to the locks along the Canal had to be sufficient to operate the capstans manually to open and close the heavy lock-gates, and judiciously mixed in skills to ensure that each district could undertake its own maintenance and repairs. Broadly, this was achieved by 1826 when there were thirty-nine workers on the establishment and twelve supernumeraries. Knowing that the Canal's income did not match its running expenses, James Davidson had to consider which workers were not really needed, or not needed in winter when the steamboats were not operating. He reckoned the Canal could function with ten men fewer, except in summer but noted, in a kindly touch, that two of these, David Duff and Kenneth Bethune who were labourers on the Canal banks, 'are actually worn out in the service and if paid off will be miserable'. His own contribution to economy was to sell his horse and buy 'a small poney' in the hope of cutting this expense by half.

In 1825 tonnage rates were doubled to ½d per ton per mile, mainly to increase income and partly to meet complaints about unfair

Opposite: Son of Matthew, James Davidson was the first Engineer supervising the whole Canal, and latterly the oldest serving Engineer in its history.

Night scene of Tomnahurich Bridge, Inverness, and the bridge-keeper's house, built in 1831.

competition from the proprietors of the Forth and Clyde Canal. 'It seems', wrote Telford, 'that the Forth and Clyde are beginning to exchange former Contempt for Alarm respecting the Caledonian Canal'. Income proved to be no higher. Many skippers were prepared to risk the terrors of the Pentland Firth to save 2/7d per ton and the rates were lowered to ¼d per ton per mile two years later.

The Canal depended on a good head of water and 1824 proved to be a summer of drought, Loch Oich's level falling some three and a half feet. Steps were taken to clear away rock where the River Garry flowed out of Loch Garry to increase the flow and raise Loch Oich. Then further up at the outlet of Loch Quoich a regulating sluice was constructed to make its water available to the Canal in times of scarcity.

Elsewhere, at Dunaincroy between Muirtown and Dochgarroch the problem of severe leakage had again to be faced in spite of earlier attempts to solve it. Although practically the whole Canal was at least fifteen feet deep by 1825, it was impossible at this place to maintain even thirteen feet. Further dredging was carried out. The water was then drawn off and a stretch of five hundred and twenty yards was laid with matting on the bed and webs of woollen cloth up the slopes to a height of nine feet. This huge woollen patch was completely covered with a thick layer of puddled clay and sand. This was William Hughes's last undertaking on the Canal before he returned to his native Wales. The repair was costly but it worked and brought this section to the same depth as the rest of the Canal.

Almost all the Canal's early customers were vessels of less than 100 tons, carrying timber, grain, slates, salt and herring. Andrew May, the Collector at Clachnaharry, expressed his excitement in a letter to Telford on 31 October 1823, when four brigs and three sloops passed through Clachnaharry before breakfast laden with herring for the Irish ports. The following year thirty ships went through in two days with the same cargo for the same destination. The Irish press noted the convenience of the Caledonian Canal, as the *Belfast Commercial Chronicle* reported on 13 August 1824:

> The smack *Margaret*, Captain Wilson, arrived with the first cargo of herrings this season from Cromarty; when the *Margaret* left that place on the 1st inst., the fishing was very good. Captain Wilson came and went through the Caledonian Canal, a distance of 62 miles, taking only 48 hours each time to pass; he says that the attention and dispatch given at the locks etc. merit the highest commendation.

By the 17th the *Margaret* passed through Clachnaharry for Cromarty, having completed her third voyage through the Canal that month. Trade was developing but without depth the Canal would not attract vessels in the American or Baltic trade, and without their custom, it would not pay. The anomaly of charging 10/- only for steamboats was shown when two large steamboats, the *Eclipse* and the *Superb*, came through from the west in February 1825. The *Superb*, a 300 ton vessel for the London to Leith run, measured 135 ft x 39 ft, just narrow enough to pass the locks, and, powered by two 55 hp engines, was capable of 10 mph. If this ship was the shape of things to come, would the Canal banks stand up to its wash? More important, would steamships become so big that the Caledonian Canal, so often criticised for being on too grand a scale, be found to be too small for them?

Even Telford on his tour of inspection in 1828 set off in a steamboat. 'Tomorrow I proceed in full staff viz. Davidson, Mitchell, Horses,

Working the locks in the old way, six men on the capstan bars, compared with one opposite.

Carriage in a Steam-Boat. G. May, who is now with me, says the trade is getting better on the Canal, and the steam-boats are moving in all quarters, between and along the East and West coasts in all directions.' Telford's example was followed by the carrier who delivered goods from Inverness to Tain. On the first voyage of the steamer *Highland Chieftain* through the Canal in 1831, he embarked at Muirtown along with his horse and cart fully loaded and steamed to Invergordon, which seemed to him the easiest way of returning to Tain. By 1834 two companies were in competition on the Inverness to Glasgow run and the *Glenalbyn* had begun a service between Liverpool and Inverness.

For the people of Inverness these services marked the beginning of the habit of travelling as well as an influx of visitors who 'overwhelmed their Inns and Lodging houses'. Beside the Canal the Muirtown Hotel, 'a tolerably respectable Inn', was providing accommodation for some of the ship-masters and passengers, and the town was becoming a centre to which the Canal was attracting travellers.

James Davidson's health gave way and he departed on half salary for the more sympathetic climate of the south of England in 1829. George May took his place but the Davidson connection with the Canal was far from ended. After some years as an engineer in Leicestershire he was

Redundant capstan and bars on the left, new control pedestal on the right for operating lock-gates.

able to return to the Caledonian Canal as Collector at Clachnaharry. As late as 1867 he succeeded George May as Engineer and held that position for the next ten years. His son, John G. Davidson, became Collector of Tolls and after a period in charge of the Crinan Canal, John was Engineer of the Caledonian from 1885 until 1912. Telford had recommended the promotions of James Davidson and George May, both of whom had been his pupils. When the Commissioners adopted the policy of promoting from among their own employees, it is not surprising that their Engineers in the nineteenth century came from only three families, all connected with the building of the Canal – Davidson (Matthew, James and John), May (George, the son of Andrew, the first Collector at Clachnaharry) and Rhodes (William, the nephew of Thomas, the Hull carpenter who had constructed the lock-gates). They were the inheritors of the Telford tradition in the north and in the case of the Davidsons, almost an engineering dynasty.

George May had lived in Telford's house in Abingdon Street in London as one of his assistants at the same time as Joseph Mitchell who succeeded his father, John, as Inspector of Roads and later wrote *Reminiscences of My Life in the Highlands*. George had been a student at Aberdeen University for three years before succeeding his father as Collector of Tolls. When he became Engineer in 1829 one of his first actions was to fix 1/- per mile as the maximum that might be charged by trackers for towing sailing vessels. He wanted to put a stop to overcharging by trackers, which antagonised ship-masters at a time when Canal dues had been purposely reduced to attract more vessels to the Canal. Then he took the novel step of having hand-bills printed by the *Inverness Courier* and distributed to the main ports. One of them is the illustration on page xiv. The cost of this exercise in publicity he dismissed as trifling: a single extra small boat passing through the Canal would pay the cost!

11
RECONSTRUCTING THE CANAL – IN THE 1840s

In November 1834 storms of heavy rain raised the level of the water in the Lochs to such an extent that the Canal banks in the west were in danger of being destroyed, flooding all the farmlands between Loch Lochy and the sea. The new Mucomer river channel was choked and Loch Lochy rose until it was three feet higher than the lock-gates at Gairlochy. Sailors from the Liverpool and Glasgow steamboats helped the lock-keepers for two days and two nights to build dams of turf and clay to try to direct the water into the reach below. The large Strone sluice for draining the Canal was opened full and all the lock-gate sluices from Gairlochy to the sea were opened to let off the flood waters. The exertions of the lock-keepers and the sailors saved the people of the Corpach valley but the flood was a frightening warning. The people who lived west of Loch Lochy were dependent for their security on a single lock at Gairlochy and a clear run-off for the River Lochy at Mucomer.

George May set about the task of repair with vigour, attending to the stonework of Gairlochy Lock and raising the Canal banks above it. He set the dredger to work in the Mucomer river channel. Part of the puddle lining near Shangan Aqueduct was renewed and the gravel which had been swept into the reach was removed by dredging.

At Fort Augustus three years later when the gates were open to allow some vessels to pass upwards, the recess wall of the lowest lock collapsed. All the masonry and rubbish tumbled into the lock forcing the gate-leaf two-thirds shut. A passage twenty-six feet wide was open but only for small ships drawing eight feet of water. Five of the eight bigger vessels in the Canal at the time found their passage blocked. The 54 ton *Catherine* of Dundee had to turn back westwards to deliver her general cargo from Liverpool to Dundee. The *Leda*, a 172 ton vessel carrying oak and bark from Bremen to the Isle of Man,

the Dundee *Eliza* with linseed from Riga for Newry in Ireland, the Stornoway *John and Thomas* with timber from Memel for Dublin and the 140 ton Danish *Thorvalo* carrying flour from Copenhagen to Liverpool were trapped in Loch Ness and had to turn back to the open sea. This was a tragedy for the Canal. Seldom had so many big trading ships been in transit at the same time and a 260 ton Russian barque, which would have been the biggest sailing ship to pass through the Canal, was approaching Clachnaharry. News that the Canal was blocked must have spread quickly to the ports of Britain and northern Europe.

In fact, most of the rubble was soon cleared out of the lock, in spite of the lack of a diving bell. The hole in the wall was blocked with turves and sand contained in fishing net and canvas bags, and held in by props of timber. By this temporary repair, the Canal was made serviceable again within three weeks.

Just before this disaster occurred, George May had been required to submit a long report on the Canal to the Secretary of the Canal Commissioners. He began by expressing admiration for the vision of Thomas Telford:

> The idea of constructing a canal on so stupendous a scale was characteristic of the bold and original genius of its author . . . had this great work been completed in the manner then proposed, or had the execution of its details at all corresponded to the magnitude and excellence of the design, it would undoubtedly have formed one of the noblest monuments on record of national skill, enterprise and magnificence.

A canal of much more modest breadth and depth would have been ample to coastal and fishing vessels. For foreign traders, however, he maintained that there should have been provision for towpaths along the lochs as well as along the cuttings, because the prevailing wind in the Great Glen, favourable to sailing ships passing in one direction was bound to detain ships trying to make headway in the opposite direction in narrow Lochs. This defect could be remedied, thanks to the development of steamships, by providing steam-tugs to tow sailing ships against contrary winds.

Opposite: George May, driving force towards the reconstruction of the Canal in the 1830s and 1840s.

That the Canal as it was built did not measure up to its grand design had been the result of mounting opposition to further Government expenditure on a project which had cost far more than had been estimated. Speedy completion at all costs replaced durability as the aim and the premature opening of the partially completed Canal had been responsible for the series of defects which had been revealed since then.

In criticising some of the masonry at the west end May used words which were not in Thomas Telford's dictionary:

> The masonry of the whole structure, judged with reference to the purposes for which it was intended, I cannot characterise by any other term than that of execrable; a worse piece of masonry than the Banavie and Corpach locks exhibit is not to be found in connection with any public work in the kingdom.

He also made two astonishing accusations: that the contractor had believed that the Canal works would never be tested by being brought into operation for purposes of navigation, and that he had taken the utmost steps to conceal the shoddy workmanship from Thomas Telford. He expressed surprise that Telford had been taken in!

It is probable that George May exaggerated to make his case convincing but this report, followed by breaches in the masonry, at Fort Augustus and then at Banavie, brought the Commissioners to the moment of decision about the Canal – should it be closed for major overhaul to bring it up to the standard that Jessop and Telford had intended or should it be abandoned altogether? Governments seldom make such decisions without advice and they called for an independent survey by James Walker, President of the Institution of Civil Engineers.

His report supported George May's observations closely but in more moderate terms. First he recommended certain measures which were essential to prevent the Canal flooding – an extra lock at Gairlochy, repair of the lowest lock at Fort Augustus, and the alteration of the waste weir at Loch Oich at a total cost of £25,000. To bring the Canal up to the standard it ought to have reached before it was opened would, he thought, cost a further £104,000. This sum would cover the repair of locks from Corpach to Fort Augustus, dredging the Canal between Laggan and Loch Oich, dealing with leaks between Dochgarroch

The pound at Gairlochy and the new upper lock in the distance constructed to protect against Loch Lochy flooding.

and Muirtown, and facing the inner Canal banks with stones to protect them against the wash from steamboats. The depth would be seventeen feet throughout, not twenty as originally intended, and deep enough to take vessels of 400–500 tons then engaged in the Baltic and American trade. Adding £14,500 for the provision of steam-tugs, he estimated the total cost at £144,000. He advised that the whole repair and improvement of the Canal be undertaken, believing that the Canal had never yet had a chance and that its prospects were now far more promising because of the coming of steam.

The matter was considered in great detail by a Select Committee of the House of Commons in 1839, which examined many witnesses, engineers like James Walker and George May, as well as shipowners and representatives of trading interests. While these investigations were going on, it is interesting that, in spite of recent accidents and temporary repairs on the Canal, more and more ships were using it: five hundred and fifty-four passed through from sea to sea in 1839, seventy-eight more than in any previous year. Here was better evidence than any words that the Canal was fulfilling a need and that money spent on its reconstruction would be justified.

In the meantime, more lock-walls collapsed, at Banavie and Corpach, as if to confirm the creaking condition of the masonry and the need for a speedy decision. A private company which was formed in Edinburgh showed an interest in leasing the Canal from the Government but they would not accept the Select Committee's terms and the Commissioners were unable to shelve the problem of what to do about the Canal.

The Treasury, however, were not willing to put up the money without first making their own enquiries about the viability of a reconstructed Canal. They appointed Captain Sir William Edward Parry RN to report from the navigation and traffic points of view. He recommended the provision of steam-tugs, an ice-breaker, and for the approaches to the Canal, some small lighthouses and beacons and the publication of correct charts. He travelled to seven major ports from Hull in the east northwards and round to Liverpool in the west to question no less than a hundred and five shipowners, masters and merchants on the value of the Caledonian Canal, past and projected, to their trade. The replies he received revealed complaints about the state of the Canal, about vessels trapped and delayed there because of adverse winds, and about sailors having to open and shut the lock-gates themselves because there was no one at the locks to lend a hand. Questions about how the coming of railways would affect the use of the Canal brought conflicting answers. In Hull a shipbroker noted how cheaply goods were now being carried by rail between Hull and Liverpool compared with the costly northern route through the Caledonian Canal. Some Leith merchants were looking forward to the opening of the railway to Glasgow to transport goods quickly to the west coast, but one of them thought the railway would not alter trade, 'We can now send goods from here to Glasgow by cart for 25/- a ton in 24 hours, and by the Forth Canal at a little slower rate, for 21/-. This is expeditious enough for the general purposes of Trade.' One of the Newcastle merchants interviewed, them engaged in the Mediterranean and West Indian trade, and formerly in the coastal trade in the Moray Firth, knew the Canal well because he had lived in Inverness for a considerable time. He was Thomas Cargill, the son of the masonry contractor at Clachnaharry. He represented the majority view, calling for the deepening and repair of the Canal and the provision of steam-tugs. In these terms, Captain Parry reported, confirming James Walker's recommendations of 1839.

At last, six years after George May's devastating report, Parliament agreed to the expenditure. James Walker drew up detailed specifications and invited tenders from contractors to undertake the whole reconstruction. The work went to Jackson & Bean of Aston near Birmingham, who began operations in September 1843 on a lump contract worth £136,089 and agreed to finish it within three years. Whereas the original construction under Telford had been done by fixing rates per cubic yard of soil removed or masonry built and paying contractors for what they had done, Walker's method required careful plans and specifications beforehand and a single contractor at a fixed price for the whole contract.

Much contributory work was done by the Canal workmen themselves. They overhauled the dredger, built new barges, carried stone and other materials to the places where they would be needed and made wheelbarrows which were sold to the contractor. Some of the Canal employees were appointed to superintend and inspect the reconstruction but their biggest undertaking was to deal with a further failure on the Canal, the collapse of the Upper Banavie culvert in March 1843. The waters of the six mile stretch from Gairlochy to Banavie rushed through the gap and flooded the surrounding land. Once the stream which used to flow through the culvert was diverted to another one, the Canal was repaired and made ready to allow materials to be carried along it for reconstruction work inland.

The accident at Upper Banavie prompted an addition to the contract, the dismantling and rebuilding of the aqueducts at Muirshearlich, Loy and Shangan at a further cost of £13,711. The contractors suffered a blow when one of the partners, Mr Bean, was killed when his horse took fright and backed over a bridge at Moy, which at that time had no parapet. In spite of this the work was carried on relentlessly by a work-force which reached fifteen hundred by the middle of 1844. Most of the tradesmen were from Inverness and Moray and most of the labourers were Highlanders. To help them they had a hundred horses and an impressive array of machinery.

The repair work began where the masonry was worst – in the west. The wharf-wall in Corpach basin was extended by a hundred

The only remaining engine house, at Fort Augustus, used as a store after 1887 when the engines were scrapped.

feet and a wharf was constructed at the top of Banavie locks. The recess walls for the lock-gates at Corpach and Banavie were taken down and completely rebuilt and the defective masonry in the lock-walls was replaced. It proved impossible to construct a passing-place in the middle of the eight locks at Banavie, to remove the Canal's bottleneck, as had been intended. Instead the contractors so improved the sluices and the capstans for working the lock-gates that eight to ten minutes were taken off the time to pass through one lock, making a saving of three hours in passing through the entire Canal. All the lock-gates were repaired and fitted with larger rollers to make them easier to operate.

Angling below the weir where the River Ness flows out of Loch Dochfour.

From Banavie to Gairlochy the whole line of the Canal was faced
with stone pitching and its channel was deepened. More spectacular
was the rebuilding of the aqueducts on the line and the construction
of the new lock at Gairlochy, where Redcastle freestone was used, as
in all the locks from Clachnaharry to Fort Augustus. Some extra
work was undertaken making a towpath at Muirshearlich, to provide
employment for Highlanders who were in distress because of the
potato famine in 1845.

Up in the Middle District, a new dredger supplied by Vernon of
Aberdeen reinforced the old one in making the Laggan Reach deeper
and wider. Together they gouged out the channel through a solid mass
of sand, stone and clay, as stubborn as concrete. The new dredger
then deepened the channel through Loch Oich while workmen built

a higher weir to keep up the water level in the Loch. Masonry and lock-gate repairs were carried out on the same lines as in the west.

Above the Fort Augustus Locks new puddle linings were put in to keep the bottom and the sides of the reach watertight. The Canal was widened to make a turning place for steamers and a wharf was also constructed. The locks required most attention. They had to be stripped down and rebuilt with Redcastle or Tarradale stone. Workmen started at the top lock and worked down towards the lowest, paying careful attention to the new recess walls and the hollow quoins which would take the weight of the lock-gates. Work on the lowest lock again proved to be the most difficult and required much pumping by steam engines, because of the amount of water coming through the old walls. Some lock-gates were completely renewed and the others repaired, and all were adapted for easier working.

At the east end of Loch Ness the entrance at Bona was made wider. A track was formed at the north end of Loch Dochfour for use by tracking horses and then it was decided to widen it to turn it into a turnpike road on the line of the present road. The weir was extended to raise the levels of Loch Dochfour and Loch Ness. All along the reach from Dochgarroch lock the banks at water level were lined with stone and the tracks laid with gravel. New puddle linings were put in on the Muirtown Reach. Above the Muirtown locks, near Burnfoot, a turning place was made for steamboats. Below the locks a new cast-iron swing bridge, which had originally been intended for Gairlochy, was installed to cope with heavier traffic on the north road, and it was the old Muirtown bridge that went to Gairlochy.

To guide mariners at the approaches to the Canal the Commissioners of Northern Lighthouses agreed to provide lights in Loch Eil and the Moray Firth. Small cast-iron lighthouses were placed at points between the Canal and the Lochs, at Fort Augustus, for example. New Admiralty charts of the Firths and the Lochs were now available. Above all, four steam-tugs, *Hero*, *Speaker*, *Secretary*, and *Engineer*, were available to tow vessels along the Canal. The Canal was complete, and ready to welcome all vessels of up to 400 tons, drawing up to sixteen feet of water. It was at last, in 1847, as it should have been in 1822 when it was opened, with the added facility of tugs.

Because its success was more predictable, this reconstruction may appear to have been less heroic than the original project to build the

Canal through the Great Glen but the job was thoroughly done and finished inside the scheduled time. Broadly, it confirmed the Canal in the shape and structures it has now and the Engineer's guide to the Canal today is an updated copy of Morrison's Survey, which was made at the time of the reconstruction.

No ceremony graced the re-opening of the Canal on 1 May 1847 but on 16 September that year Prince Albert passed along it on board the steam yacht, *Fairy*. This 317-ton screw-steamer taught George May a lesson, because on approaching a lock it could not stop as quickly as a paddle-steamer. Sensing the danger to the works, he determined to charge steamships with propellers almost as much as paddle-steamers in future. Prince Albert went on to Dochfour and then to a rapturous welcome in Inverness where he attended the Northern Meeting Ball, unaware of the financial consequences of the first royal progress through the Canal.

Heavy guard chains were also installed above and below each single lock and flight of locks to protect the gates against steamships which failed to stop quickly enough. The chains were lowered by a windlass to allow a ship to pass through and raised to the guard position again immediately afterwards. Some of the time saved on the improved capstans was lost on the windlasses. This extra duty for the lock-keepers began soon after *Fairy*'s voyage and lasted for more than a century.

The new works were tested by the great flood of January 1849 which burst through the Canal bank at Dochgarroch and also swept away the old seven-arched bridge over the Ness at Inverness. Elsewhere the Canal stood the strain. In 1850 George May went off on a tour of the north Baltic ports such as Riga to tell owners and masters about the facilities and safety of the reconstructed Canal compared with the Pentland Firth. The trip must have helped because the number of vessels using the Canal on voyages to and from foreign ports doubled in a year to seventy-one, with a registered tonnage of 10,000 tons. The next year he was off again, equipped with printed notices, plans and new Admiralty charts to distribute in Oslo, Copenhagen, Stockholm, Hamburg and Stettin. If he could attract the big Baltic traders he was sure the Canal would pay but soon the Crimean War with Russia put a stop to the Baltic trade.

12

WORKING THE CANAL – AFTER FIFTY YEARS

EVEN the reconstructed Canal, now capable of taking ships with a draught of sixteen feet and equipped to tow them against adverse winds, was not an economic proposition, although in the 1860s it was just paying its way. Passenger steamers were as busy as ever but the trade in herring and potatoes to Ireland and the West of England began to decline. Nor had the Canal stimulated the growth of industry in the Great Glen, as had been hoped. The transition from sailing ships to steamers reduced the demand for the services of the tugs on the Canal while the increase in the size of steamships meant that many could not pass through the Canal. In any case, bigger steamers had far less difficulty than sailing vessels in negotiating the Pentland Firth. As a result, the Baltic traders did not become the economic saviours of the Canal, as George May had hoped.

The period of the American Civil War (1861–5), however, helped to increase the number of through voyages. Cut off from the usual sources of raw cotton in America, some West of Scotland manufacturers turned again to linen production, using flax from the Baltic. In addition, considerable cargoes of iron passed through the Canal for railway-building in the north. In 1866 the total number of voyages through the Canal by sailing ships and steamships carrying passengers and cargo was over two thousand.

On the death of George May, aspiring successors were busily manoeuvring like bees round a honeypot. A letter written then revealed that their search for referees and testimonials was in vain: 'One of the existing staff on the Caledonian Canal, who formerly

Opposite: An optimistic map showing trade of the western world passing through the Canal after reconstruction, 1852.

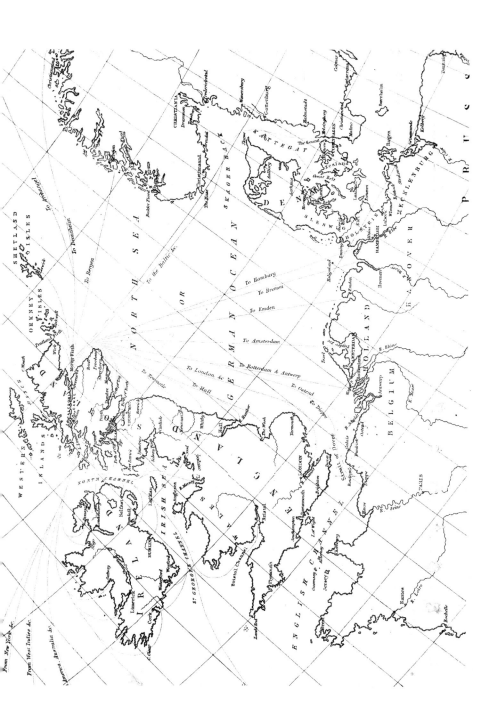

filled the office, and has lately been acting as Collector, has now taken both positions'.

This was James Davidson: he was then sixty-nine. His son, John, became his assistant and Collector, John living in the Engineer's House at Clachnaharry, his father in Burnfoot at the top of Muirtown Locks, where he also had some land. This was not to be a period of vigorous development but of routine operation and care and maintenance of an established Canal.

He revised the Table of Rates for reprinting in 1871. Tonnage dues remained remarkably low. On unladen sailing ships the charge was only a shilling per registered ton for the whole passage, or ⅓d per ton per mile, and on loaded ships ⅓d applied at 6d per ton on the ship and 9d per ton on the cargo. Pleasure yachts paid 1/6d per ton, steamers with screw propellers 1/9d and paddle-steamers 2/-, all for the full passage of the Canal. Towage by steam-tug cost 6d per ton for laden vessels through Loch Ness, 2d through Loch Oich and 3d through Loch Lochy. Trackage rates for each horse ranged between 6d and 1/- per mile by agreement but 1/- was still the maximum that could be charged.

Muirtown Basin was open to ships with a draught of eighteen feet and Corpach to seventeen feet. The Table enumerated the dues to be paid for loading and unloading different cargoes in the basins and at piers and jetties along the Canal. As examples, the rate for coal was 3d per ton and for peats 2d per 1,000; for potatoes 4d per ton and for herring 2d per barrel; for building stone 2d per ton and for slates 6d per 1,200; for bones and guano 8d per ton, kelp 4d and common manure 2d. The list was long and it was useful to have it in booklet form for the convenience of ship-masters.

1872 marked the fiftieth anniversary of the opening and the twenty-fifth of the reopening of the Canal after its reconstruction but they passed without notice. Instead, James Davidson found himself fighting off criticism. In November he was writing to the *Inverness Courier*, denying that the Canal works were a cause of recent floods in Inverness, and dealing savagely with the allegation in a Mr Bidder's Report that the sill of the lowest lock at Fort Augustus has been built six feet higher than it should have been. He pointed out that the foundation of this lock had been constructed under great difficulties, twenty-four feet below the water-level of Loch Ness and admitted that

Gondolier at the top of Muirtown, Inverness.

it was too high, thirteen inches too high. He wrote with conviction and authority: he had been there when the lock was built, fifty-five years previously.

He had to face more accusations about the Canal and its management from Kennedy McNab of Inverness who in 1876 addressed himself to Sir Erskine May, Speaker of the House of Commons and Chairman of the Canal Commissioners. When McNab gained no satisfaction he expressed the view that the Canal should be removed from the control of the Canal Commissioners altogether.

Demands for more pay by the workmen in 1872 marked the beginnings of primitive collective bargaining by the lock-keepers. Because the workers were scattered along the length of the Canal it must have been difficult to organise them but a meeting was held, chaired by James Bain, carpenter and storeman, and the sum of 3d per man was levied to pay a letter-writer to advocate their cause. The wages movement was fairly successful in 1872 but the increases were smaller for the skilled workers who, according to Davidson, were better off than corresponding tradesmen on the railway, because of their additional perquisites, free houses, gardens and gas.

SUMMER TOURS IN SCOTLAND.

THE ROYAL ROUTE.

Glasgow and the Highlands

Via Crinan and Caledonian Canals.

TOURISTS' SPECIAL Cabin Tickets, issued during the Season, Valid for *SIX SEPARATE* or consecutive *DAYS'* sailing by any of Mr. David MacBrayne's Steamers,

£3.

ROYAL MAIL STEAMERS.

COLUMBA,	PIONEER,	LINNET,	ETHEL,
IONA,	LOCHAWE,	ISLAY,	
CHEVALIER,	GLENCOE,	HANDA,	
GRENADIER,	CLAYMORE,	LOCHIEL,	
GONDOLIER,	CLANSMAN,	FINGAL,	
GLENGARRY,	CLYDESDALE,	LOCHNESS,	
MOUNTAINEER,	CAVALIER,	INVERARAY CASTLE.	

The Royal Mail Swift Passenger Steamer

"Columba" or "Iona"

Sails daily from May till October, from Glasgow at 7 a.m., and from Greenock about 9 a.m., in connection with Express Trains from London and the South, Edinburgh, and Glasgow, etc., for KYLES OF BUTE, TARBERT, and ARDRISHAIG, conveying Passengers from OBAN, GLENCOE, INVERNESS, LOCHAWE, STAFFA and IONA, MULL, SKYE, GARELOCH, STORNOWAY, THURSO, etc., etc.

Tours to the West Highlands,

OCCUPYING ABOUT A WEEK,

BY STEAMSHIP

"Claymore" or "Clansman,"

Via Mull of Kintyre, going and returning through the Sounds of Jura, Mull, and Skye, calling at Oban, Tobermory, Portree, STORNOWAY, and intermediate places.

CABIN RETURN FARE, with superior Sleeping Accommodation, 45s.;

Or INCLUDING MEALS, 80s.

The Route is through scenery rich in historical interest and unequalled for grandeur and variety. These vessels leave *GLASGOW* every Monday and Thursday about 12 noon, and Greenock about 5 p.m., returning from Stornoway every Monday and Wednesday.

The Steam-ship CAVALIER

Will leave Glasgow every Monday at 11 a.m., and Greenock at 4 p.m., for Inverness and Back (via Mull of Kintyre), leaving Inverness every Thursday morning. Cabin Fare for the Trip, with First-class Sleeping Accommodation, 30s.; or including Meals, 60s.

Official Guide Book, 3d.; Illustrated, 6d.; Cloth Gilt, 1s. Time Bill, Map, and List of Fares,

Sent free on application to the Owner,

DAVID MACBRAYNE, 119 HOPE STREET, GLASGOW.

MacBrayne's summer tours.

The outstanding social occasion in the history of the Canal took place on 16 September 1873 when Queen Victoria travelled along it from Banavie to Dochgarroch in the *Gondolier*. She showed more concern for anniversaries than the Canal authorities, for she recalled that Prince Albert had sailed on the Caledonian Canal twenty-six years earlier, also on 16 September. She was impressed: 'The Caledonian Canal is a very wonderful piece of engineering,' but added: 'Travelling on it is very tedious.' At the locks she was even amused: 'It was amusing to see the people, including the crew of the steamer, who went on shore to expedite the operation, run round and round to move the windlasses.' Crowds assembled to see her pass and at Fort Augustus she passed through a triumphal arch. Landing at Dochgarroch she travelled in an open coach, with an escort of the 7th Dragoon Guards, through enthusiastic crowds to Inverness station to travel by Royal Train back to Balmoral.

Gondolier, on which Queen Victoria travelled through the Canal, below the locks at Fort Augustus.

Journeys through the Canal by steamer became popular and fashionable and steamships, hotels and other attractions were in competition to cater for the trade. In his *Two Months in the Highlands, Orcadia and Skye*, Charles Richard Weld found a well-informed guide in Captain Turner of the *Edinburgh Castle II* (later the *Glengarry*). At Fort Augustus he encountered Gordon Cumming, the Lion Hunter, followed by his servant and a long-bearded goat, inviting passengers to inspect his famous museum collection at a shilling a head, while their boat negotiated the locks. Gordon Cumming himself was an attraction as unusual as his collection of skins and heads. On hot days he might appear kiltless. At other times he sported kilt and plaid of tartan with top-boots, frilled shirt and masses of jewels, a brass helmet on his head and silver fish-hooks in his ears.

Black's Guide contained advertisements about competing hotels in Inverness: Marshall's Station Hotel, 'specially built as a hotel', the

Omnibuses at Muirtown waiting for the passenger steamer.

Caledonian 'patronised yearly by the best families of Europe', as well
as the Union, White's Royal and the Waverley. Omnibuses went out
to meet all the steamers at Muirtown. Even at lonely places on the
Canal, such as Laggan, little shops were opened to cater for the needs
of tourists. For the wives of lock-keepers, however, the tourist trade
brought restrictions. When a passenger steamer was passing they had
to keep their children away from the side of the lock and they were
forbidden to have washing hanging on the line!

Langland's *Princess Beatrice* brought passengers weekly in summer
and fortnightly in winter from Liverpool to Inverness for 25/- cabin
class or 10/- steerage. It also carried cargo, and connected with English
canals, claiming to offer the cheapest rates for taking goods from the
North to places like Manchester, Birmingham and Wolverhampton.
David Hutcheson's *Gondolier* and *Glengarry* offered a daily service
along the Canal in either direction. Leaving Banavie for Inverness at
8 a.m. and collecting passengers along the way the steamers connected
with trains to the north and to Aberdeen from Inverness. Their *Lochiel*
and *Lochness* operated between Inverness and Fort Augustus delivering

and collecting mail as well as passengers at Temple Pier, Inverfarigaig, Foyers, Invermoriston and Fort Augustus. Competing with the *Princess Beatrice*, Messrs Hutcheson's used *Staffa*, *Cygnet* and *Plover* on a twice-weekly service between Inverness and Glasgow, carrying both passengers and cargo. There were even pleasure excursions to Invermoriston on occasional Saturday afternoons on the *Glengarry*, Cabin 2/6d and Steerage 1/6d, and the chance of a drive up the Glen for seven or eight miles to admire the scenery at a extra 1/6d a head. A call at Foyers on the return journey gave time to visit the Falls before returning to Inverness by 10 p.m.

The route was attracting people who wanted to look at scenery or do something different – 'Let us go through the Caledonian Canal.' It sounded an exciting challenge. Since 1860 the Canal authorities had been making charges of 4d to 1/- on passengers in transit. One of the regular steamers, Messrs Hutcheson's *Staffa* paid £6/10/- each way and Messrs Langland's *Princess Beatrice* on the Liverpool–Dundee run gave the Canal an income of about £470 a year. But James Davidson worried more about ships which were not using the Canal. A Norwegian firm, whose ships, the *Carl* and the *Stadt*, had passed through regularly, found them too small for the trade and intended to use two larger vessels, which were too big to pass through the Canal.

The four timber-hulled tugs, purchased for the Canal in 1847, did not have a long life. The *Hero* and the *Secretary* were disposed of in 1860 to pay for repairs on the *Engineer*, and she and the *Speaker* were sufficient to deal with all the demands for towage, as steamships were gradually replacing sailing ships. Then Davidson reported that the *Speaker* was utterly worn out and that *Engineer* would serve a little longer, if repaired. The search began for a suitable vessel at a reasonable price and an order was given to Cunliffe & Dunlop of Port Glasgow for an iron screw tug. This vessel, named *Scot*, was powered by two 10 hp engines, which proved able to tow a barge 200 feet long without difficulty. Her economical running impressed James Davidson for she made the trip from Greenock to Inverness in April 1876 on only

Page 130: William Rhodes, the railway builder who became a canal engineer, first on the Crinan, then on the Caledonian.

Pages 131: John G. Davidson, the third of the remarkable Davidson family to be Engineer at Clachnaharry.

three tons of coal. Reinforced with plating all round her waterline, she worked as an ice-breaker in winter in addition to undertaking all the towing that was now required.

At 8.30 a.m. on 19 August, 1876 James Davidson dispatched a wordy telegram from Banavie to Sir Erskine May, Speaker of the House of Commons in Westminster:

> *Staffa* steamer has completely destroyed upper lock gates at Laggan Locks yesterday. Through passage of Canal quite stopped damage most serious will take at least three months to restore works will write fully on return home and more fully when damage is ascertained by diving on next Tuesday. Meantime passenger traffic is open by transfer at the Laggan Locks.

A new Russian dredger on its way to Kronstadt in the Baltic had been in the upper lock. Hutcheson's steamer, *Staffa*, was too close to the lock-gate above it. The lock-keeper ordered *Staffa* to move further back. 'Full speed back!' roared the captain. The engineer was having his dinner and an inexperienced young sailor excitedly went 'full speed ahead'. *Staffa* lunged forward, cut through the guard chain and crashed through the upper lock-gates. Fortunately she did not hit the dredger or reach the lower gates but the upper gates were shattered. This was the second time *Staffa* had been responsible for an accident on the Canal, having smashed through a gate at Banavie five years earlier.

The damage caused by this accident at Laggan seemed to be too serious for the Canal workers to repair without outside help and a Donald Cameron wrote to Sir Erskine May recommending that he call in a Glasgow firm without delay. James Davidson assessed the damage and decided that the work could and would be done by the Canal staff themselves. The level of Loch Oich being low, it was possible to operate the Canal with a single lock and swift steamers continued to pass through. Passengers on the *Gondolier* and the *Glengarry* were transferred at Laggan to another boat brought up by Messrs Hutcheson's and in this way passenger services were uninterrupted. The debris was cleared out of the lock and measurements were taken for a new lock-gate. It was built at Clachnaharry, towed up the Canal and carefully installed. Just five weeks after the accident, the Canal having been closed for only eight days, James Davidson was able to inform

Sheep being transported by boat along the Canal.

the Speaker by telegram from Laggan on 26 September that the Canal was again open: 'Fixing new gate accomplished yesterday afternoon fits and works well difficulty and inconvenience now ended'.

It was the last achievement of the old engineer. The next year he fell ill and died. Practically his whole lifespan of almost seventy-nine years had been spent on the Canal and the arrangements for his funeral were singularly appropriate. The tug *Scot* collected all the men who could be spared from the west end and brought them to Inverness. The Canal workers accompanied him from his home above Muirtown Locks on his last voyage along the Canal in the steamer *Glengarry* to Tomnahurich Cemetery, where he rests overlooking the Canal.

James Davidson's immediate successor was not his son, John, but William Rhodes, the Engineer on the Crinan Canal, which had been under the Commissioners since 1817 when Alexander Gibb of Aberdeen reconstructed it under Telford's direction. This was the first time that a Crinan Engineer had taken over the Caledonian. William Rhodes had had an interesting and varied career. He had served on steam dredgers on the Shannon under his uncle, Thomas Rhodes, and had been a superintendent with Jackson & Bean, the Birmingham firm which reconstructed the Caledonian Canal. He had worked on railway construction abroad for sixteen years, as an agent for Thomas Brassey, the great international contractor, first on the Paris to Cherbourg Railway, where Joseph Locke was engineer, and then in Spain and on the Central Argentine Railway. In eight years at Crinan he gained a reputation for getting things done and had brought the Canal to a new level of efficiency.

John Davidson took Rhodes's place at Crinan and returned to Clachnaharry in 1885 to be Engineer of the Caledonian Canal for the next twenty-seven years. L. John Groves of Crinan took over the Caledonian Canal as well in 1912, with Eustace Porter as Assistant Engineer at Clachnaharry. Rhodes concentrated on repairs to the masonry of the locks, while John Davidson undertook a programme of lock-gate renewal between 1887 and 1893. Most of the lock-gates were over seventy years old, each pair made of oak and cast-iron, weighing about 42 tons. The new gates of oak and steel were lighter and easier for the men to handle. Except for these works, the Canal remained open, and almost unchanged, providing a service for all who wanted to use it down to the First World War.

13
WORKING THE CANAL –
AFTER A HUNDRED YEARS

IN 1920 the Caledonian Canal Commissioners handed over the Canal to the Ministry of Transport. They probably did so willingly. For a century they had been nominally responsible for it without ever having much control, since they had no expert knowledge and the Canal was far away. Their principal role, always an uncomfortable one, had been to request supplementary grants from Parliament whenever Highland floods or erratic steamers did serious damage to the works. In several years when nothing happened to it, the Canal had operated at a modest profit but never accumulated a large enough surplus to pay for major repairs. There had been times, for example in 1840 and again in 1860, when the Commissioners tried to hand it over to private enterprise but with no success.

Just as its future had been in doubt in the early 1840s when Parliament had had to choose between costly restoration and closure, the fate of the Caledonian and all other canals was considered at length by the Royal Commission on Canals and Waterways from 1906 to 1908. It was a time when the locomotive was king and canals were in decline. The question was: if canals did not pay, should the public be expected to support them out of taxation? Supporters of railways protested that even railways paid taxes which would be used to subsidise their competitors, the inland waterways. On the other hand, the canal interests argued that railways were transporting goods below cost in many instances. Later Eustace Porter, as Caledonian Canal Engineer, complained of the wear and tear on bridges maintained by the Canal, by carts taking timber to stations for transportation by rail, cargoes which ought to have been carried on the Canal.

It was an opportunity for criticism of the Caledonian Canal which was considered to be too small, in contrast to the comments which had been made about its size when it was under construction. Various proposals were made to enlarge and modernise it, in the hope of attracting industry to the Great Glen but John G. Davidson maintained that vessels were now navigating the Pentland Firth without much difficulty and could not be won back, even to an enlarged Canal. The Royal Commission was unmoved by plans for modernisation of the Caledonian Canal and did not recommend public expenditure for that purpose.

During the First World War, ships continued to pass through the Canal in spite of the withdrawal of almost all passenger services, the requisitioning of large numbers of trawlers and drifters for service with the Royal Navy, and the reduction in the Canal staff as men departed on active service. The mood of the annual reports by the Commissioners was depressing on account of the drop in income. In fact, the record of the Canal in these four years is a proud one. Only once, in 1915–16, did the number of passages through the Canal fall below two thousand a year.

Fishing boats steamed out to sea as usual and when the east coast fishing was closed in autumn 1916, the fishermen passed through the Canal to the west to fish there. Many vessels on Government service used the Canal, five hundred and forty-one in the year to the end of April 1918, and the amazing number of 3,715 out of a record total of 5,439 passages to April 1919. Lock-keepers gave up their normal hours of sunrise to sunset to work on night-shift and on Sundays as well, to cope with the queues of ships passing through the locks. The reason for this traffic on such an unprecedented scale was the carriage of components for mines and other equipment from the west to the new American base at Muirtown. There, mines were assembled and tested before being taken out and laid in a vast minefield stretching from Orkney to the coast of Norway, to keep German submarines out of the Atlantic. This American enterprise was recalled in a letter in the *Inverness Courier* on 31 December 1971 by Mr E.P. Morley of Butler, Tennessee, who had been a mine-planter on board *USS Canandaigua* stationed in Inverness in 1918. Many of the vessels which conveyed these cargoes through the Canal were from the Great Lakes and after the War, men who volunteered to crew them back to Canada were allowed to settle there.

No charge could be made for these passages on Government service.

This explains the gloom in the annual reports, because their value in 1918–19 would have been £13,000, if they had been charged at ordinary rates. What is far more important is that the Canal was being of use, in the way Thomas Telford had intended, as a safe passage in time of war for naval vessels, admittedly small ones by Grand Fleet standards. The value of the Canal route in saving men's lives and contributing to vital naval strategy during this period of the War must be weighed against the charge that the Canal was a 'white elephant', when judged in purely commercial terms.

This traffic had imposed an enormous strain on both the Canal staff and the Canal works. There had been almost no opportunity to carry out maintenance work at all and the Canal was therefore closed in the summer of 1920. The masonry of locks at Corpach and Banavie was put in order but not enough money was available to overhaul those locks which required the construction of coffer-dams before they could be drained of water. The repairs cost more than usual because they had been postponed so long. In the last report they presented, the Commissioners were concerned even more because normal expenditure had more than doubled, on account of higher wages and prices. The excess of expenditure over income was £13,000 – a problem for the Ministry of Transport and the beginning of a new pattern in the Canal's financial affairs.

The centenary of the Canal's opening passed without ceremony in 1922, except that the marble plaque, engraved with Southey's tribute to Telford's works, was discovered and erected on the wall of the Canal Office at Clachnaharry. Having no present occasion to report, the *Inverness Courier* reprinted its account of the opening in October 1822, which is quoted in Chapter 1. The *Glasgow Herald* noted the centenary in its Casual Column, emphasising the expense of building the Canal up to 1822 and reconstructing it in the 1840s. The repairs in 1920, which required the east end to be closed for a month and the through passage for two, were described as 'another costly overhaul [which] had to be undertaken' and the conclusion was drawn that 'on the whole it is questionable whether the Canal has ever justified the expenditure it has entailed'.

Two plans drawn up at the time by the Engineer, L. John Groves, were attempts to bring the Canal into the twentieth century. The more ambitious of the two would have made Loch Ness the summit level

Repairs at Fort Augustus in 1920 showing in the foreground the wide V-shaped sill against which the gates close and the chain for closing the left gate-leaf.

of the Canal, at half the height of the present summit at Loch Oich. The number of locks would have been reduced from twenty-nine to six, each 500 feet by 60 feet by 26 feet. Between the three locks at each end, which were big enough to take all coastal and Baltic traders, there would be an uninterrupted waterway for forty-seven miles. Such a canal would require fewer men to operate it, would be easy to maintain and would pass ships through much more quickly. Alternatively, if Loch Lochy became the summit, ten locks would be required, each 350 feet by 55 feet. Neither scheme was costed

Fishing boats above Muirtown as far as the eye can see, the paddle-steamers at rest beyond.

and neither was adopted. These plans were Mr Groves' last official thoughts on the Caledonian Canal and he was succeeded in 1921 by Eustace Porter, who had been his assistant at Clachnaharry.

Instead, the Ministry of Transport turned its attention to trying to make the existing Canal pay. Charges were raised, some by 100 per cent, with the result that traffic was slow to return to pre-war levels. Goods and passengers from Glasgow had transferred to the railways in wartime and could not be won back, with the result that the service from Glasgow to Inverness ended in 1927. Local travellers between Spean Bridge and Fort Augustus found the train quicker and cheaper. In 1924–5 the number of through and partial passages exceeded two thousand but that figure by itself gives a misleading impression of activity on the Canal. Some six hundred of these partial passages were no more than the sum of the daily return trips by the mail steamer between Fort Augustus and Inverness, passing through only one of the Canal locks at Dochgarroch. 1927–8 was the year when the drifters returned, making 1,492 passages, more than half the total of 2,641. There were also three hundred and forty-five voyages by cargo vessels and seven hundred and fifty-one by passenger steamers. August was the boom month for passengers, when on ninety-three passages nearly nine thousand passengers were carried, compared with fifty passages and only ninety-six passengers in the whole month of February 1928. Twenty thousand passengers travelled on the Canal that year but within two years the number declined to almost half, due mainly to competition from motor buses.

MacBrayne's ships on the Canal in the Twenties were the veteran paddle-steamers, *Glengarry* and *Gondolier*. *Glengarry* had been on the Canal since 1846 under the name *Edinburgh Castle II* until she was refitted and re-christened in 1875. *Gondolier*, on which Queen Victoria had travelled on her journey through the Canal, had been specially designed in 1866 for passing through Canal locks. For years they had operated a daily service between Banavie and Inverness, until the *Gairlochy* became *Gondolier*'s partner. The *Gairlochy* was destroyed by fire at Fort Augustus in 1919 and her remains can still be seen from the pier. After this, *Gondolier* worked the Banavie–Inverness run by herself in summer taking a day each way:

Monday, Wednesday & Friday		*Tuesday, Thursday & Saturday*	
Inverness	8.30 a.m.	Banavie	11.10 a.m.
Foyers	10.25 a.m.	Gairlochy	11.55 a.m.
Fort Augustus	12.00 noon	Laggan	1.20 p.m.
Cullochy	12.50 p.m.	Cullochy	2.10 p.m.
Laggan	1.45 p.m.	Fort Augustus	3.30 p.m.
Gairlochy	3.00 p.m.	Foyers	4.30 p.m.
Banavie	3.50 p.m.	Inverness	6.30 p.m.

Glengarry left Fort Augustus at six o'clock every morning, carrying mail, cargo including livestock on sale days, and passengers, and calling at piers on Loch Ness on the way to Inverness and returned in the late afternoon. After a useful life of eighty-three years *Glengarry* was withdrawn and broken up in 1927, reputedly one of the oldest steamers in the world. The Fort Augustus–Inverness run was abandoned in April 1929. Those who remember these ships plying on the Canal recall them with affection and, thinking of the bustle of passengers, luggage and coaches at the quays at Banavie and Muirtown, associate them with the great days of the Canal.

In fact, in the 1930s, the Canal was being used less and less. With only one steamer operating between Banavie and Inverness in summer, David MacBrayne Ltd put on the small passenger steamer *Princess Louise* between Inverness and Fort Augustus and carried over two thousand passengers a season from 1934. Many of them had been

What the *Girl Patricia* did at Banavie in 1929.

attracted by newspaper headlines reporting that a monster had been seen on Loch Ness. The Monster became famous in 1932 at the time of the construction of the new high-level road from Inverness to Fort Augustus. One explanation of its sudden emergence was the disturbance of the waters of the Loch by rock-falls and tree-felling while the work was in progress. Another was simply that the high road gave a better view of the Loch, especially near Urquhart Bay where most sightings have been reported. There was no mention of it at the time the Canal was constructed and it has even been suggested since that it came up the Canal from the sea. The first account of a water monster in the River Ness, not the Loch, however, dates back to the time of St Columba. That there is an unexplained 'something' in Loch

Ness cannot be denied in view of the number of sightings, especially by those local people who had previously been known to discount the existence of the Monster. It became a tourist attraction and a scientific problem which attracted many investigating teams with sophisticated equipment and could not be dismissed as local mythology or tourist gimmickry.

Fewer fishing boats were passing through the Canal, the total falling from over a thousand in 1931–2 to under four hundred in 1938–9. The official reason for the decline was accepted as depression in the fishing industry. Canal income fell by almost half in seven years and the excess of payments over receipts doubled. It is not surprising that the impression was spreading that time was running out for the Canal, that it was losing business and that if it had been in private hands it would have had to close down.

Old men used to tell the story of the accident at Banavie on 22 February 1929. A drifter, the *Girl Patricia*, crashed through both pairs of gates of the top lock, was swept through into the next lock and damaged its gates before being brought to a stop. The sudden increase in the pressure of water above the top lock made the south wall bulge dangerously. If it had collapsed, all the water in the six miles of the Banavie reach would have cascaded down and flooded the village.

When the Canal was emptied to allow the bulging masonry to be taken down it was revealed that the rubble walling behind it was not thick enough and built of stones which were far too small and irregular for sound lock construction. This discovery confirmed the criticisms made by George May about the masonry at Banavie a century earlier.

At a cost of £5,775, a new tug and ice-breaker, *Scot II,* replaced the original *Scot* in March 1931. More powerful than her predecessor, she had a specially shaped hull for breaking ice by riding over it, and was available for towing when required. Six years later when Eustace Porter retired after twenty-five years at Clachnaharry, Frank Whyte became Engineer and Manager. He took on a hundred extra men in the spring of 1939 to reinforce the Canal staff in a major overhaul of the Canal, lasting two months. The hours they worked were from 7.30 a.m. to 6 p.m. Monday to Friday with an hour off for a midday meal, and from 7.30 to noon on Saturday, making a 52-hour week. Details of their rates of pay are of interest to compare with wages paid to the men who built the Canal in the early 1800s on the one hand, and also to compare with the present day. Skilled masons employed on pointing the lock walls received 1/7d per hour, semi-skilled labourers employed pointing with them 1/3d to 1/4d per hour, or for replacing the stone pitching on the banks 1/3d per hour. General labourers were paid 1/1d per hour, the same as retired Canal workers who received £2/16/4d a week for taking charge of a gang. By the summer the Canal was in good working order. When war broke out, ships on Government service again resorted to the Canal in considerable numbers to take advantage of the safe passage from sea to sea.

14

WORKING THE CANAL – AFTER
A HUNDRED AND FIFTY YEARS

The rejection of expensive proposals to enlarge the Canal after the First World War meant that in 1972 it was still more or less as Jessop and Telford had planned it. A hundred and fifty years old, it was ancient but not abandoned, historic but not derelict. Instead it is still a working Canal, capable of passing ships of the same size as they intended, except that their draught was limited to 13½ feet due to leak-sealing over the years.

When the Ministry of Transport transferred the Canal to the British Transport Commission in 1948, having operated it, maintained it and paid its deficits annually, they had found no answer to its financial problems. Receipts were as low as they had been before the Second World War and, due to inflation, operating costs had soared. In the early 1950s, fewer vessels used it with every increase in toll charges and settled at about seven hundred and fifty passages a year. About two-thirds of these were by fishing vessels, the number of cargo vessels dropping to below a hundred and the number of yachts rising above it for the first time in 1956–7. Among the vessels OHMS which passed through in 1958 was the royal barge, in which the Queen, the Duke of Edinburgh, Prince Charles and Princess Anne sailed from Gairlochy through Loch Lochy and Loch Oich to Cullochy on 18 August, making the third royal voyage in the history of the Canal.

In their report in 1958 the Bowes Committee of Enquiry found that, even if the Canal did not pay, there was a clear case for keeping it open, on the grounds that 'the waterway is a social service'. This was new thinking, which had not been accepted at that time in the case of

Opposite: Clachnaharry sea-entrance thrusting out into the deeper waters of the Firth. Inset shows lock-keepers' houses.

uneconomic rail routes. It could not be argued that the Canal was a
social service for the people in the Great Glen, other than the men who
worked on it, because ships usually pass through the Canal rather than
serve places along it. It was, however, of great service to the inshore
fishermen. The bulk of the fishing boats were from Aberdeenshire
and Moray Firth ports and as long as they sailed to the shellfish and
herring grounds off the west coast, they would want to use the Canal
because it saved time, and fuel, and wear and tear on their boats. If the
time taken to pass through the Canal could be reduced, they would
benefit and more traffic might be attracted.

With this in mind a modernisation programme was begun in 1959,
which may prove to be as significant in giving the Canal a new lease
of life, as the reconstruction supervised by George May in the 1840s.
The intention was to mechanise all the lock-gates on the Canal. At first
it was hoped to operate the existing capstans and chains by electricity
but after trials, the hydraulic principle was preferred. The capstans
and chains were dispensed with and the gates are opened or closed by
a steel crank arm operated by a hydraulic ram mounted in the gate
recess in the wall of the lock. The controls are either in pedestals near
the lock-gates at the flights of locks, or in a control cabin at single
locks, such as Dochgarroch. All forty-two pairs of lock-gates had been
mechanised by 1968 with other associated works at a cost of £195,000.
Vessels could pass through the Canal in about two hours less than
previously. Fewer lock-keepers were needed, especially at flights of
locks. There were five lock-keepers at Muirtown in 1911, eight at Fort
Augustus and twelve at Banavie: in 1972 there were only two at each
place. Formerly lock-keepers were tradesmen and each district was
responsible for its own maintenance and minor repairs: in 1972 there
was a mobile repair crew of six tradesmen, including two divers, and
two labourers, based at Clachnaharry, who could carry out repairs
anywhere on the Canal.

For the lock-keepers the work was no longer physically hard. Lock-
gates and sluices are operated by one hand on a lever, compared with
two men making seven complete turns with capstan bars on the old
capstans on either side of the lock. Seven turns to open, seven turns
on the neighbouring capstan to close, it took seventy revolutions to
see one ship through Muirtown, eighty-four through Fort Augustus
and a hundred and twenty-six through Banavie. As a lock-keeper with

nearly forty years' service put it at Muirtown, 'When you passed ten ships through in a day in the old days, you knew you'd been working and were ready for bed.' One old capstan has been retained at each flight of locks, but the chains are not connected to the gates.

At Corpach another modernisation project began in 1964 to allow 1,000-ton ships to enter and unload timber in the basin for the new Scottish Pulp and Paper Mills, being erected nearby at Annat Point by Wiggins, Teape & Co. The sea-lock was lengthened to take vessels 203 feet long and a new wharf was constructed, complete with two Scotch derrick cranes, each capable of lifting seven tons. The basin was now better equipped but was narrower, and large vessels could not turn in it. Having unloaded, they returned to sea stern-first. The construction of the Pulp Mill brought some extra traffic to the Canal, as railway building had done earlier, none more unusual than six giant digesters built at Leven on the east coast. Each being too wide to be delivered by road or rail, they were given keels made of old railway lines to keep them steady. They were hauled by lock-keepers with ropes from Clachnaharry sea-lock to the top of Muirtown Locks and then *Scot II* towed them in pairs along the Canal. One pair like great red bubbles against the blue of Loch Ness excited considerable interest at Fort Augustus on a sunny day. Manpower replaced engines again to lower them down through Banavie Locks. The coming of the paper industry to Corpach raised hopes that the Canal would be used for transporting timber from the forests along the Great Glen, since the Pulp Mill consumed ten thousand trees a day. Road haulage triumphed, however, being more direct and requiring less handling.

As forestry did not come to the rescue of the Canal, neither did industry using hydro-electric power. The example of the aluminium industry at Foyers on Loch Ness as early as 1896 was not followed and that plant closed down in 1967. The post-war achievements of the North of Scotland Hydro-Electric Board in producing power have been paralleled by their ability to distribute it. Hydro-electricity is a form of power which is easily exportable and industries have developed elsewhere, some using power produced in the Highlands. Although classified as commercial by the British Waterways Board who became responsible for Britain's canals in 1962, the Caledonian Canal is still a waterway through mountain landscape with hardly a trace of industry along its shores. There were, however, one or two hopeful signs for the

Sailors cheerfully working the lock gates in the old way – seven revolutions to open, seven to close.

revival of trade in 1972. Paper made at Corpach was being exported to Belgium via the Canal and this trade with Europe might grow. Alumina, being imported in bulk to British Aluminium's new smelter at Invergordon on the Cromarty Firth, could be transported in smaller quantities along the Canal to their older-established smelters at Fort William and Kinlochleven. Unfortunately the Invergordon smelter was shortlived.

Fishing boats were the Canal's most faithful customers for through passages, and often sheltered at Corpach basin. In exceptional years like 1966, when Kessock herring were in abundance and Inverness harbour was crammed full of fishing boats, others tied up in the sea-lock reach at Clachnaharry. Even they, however, were using the Canal less because east coast crews fishing in the west usually left their boats there and returned home for the weekend by car. Tolls were raised in 1965 to £5 for fishing boats up to 30 ft long, and £7 for boats 30–50 ft

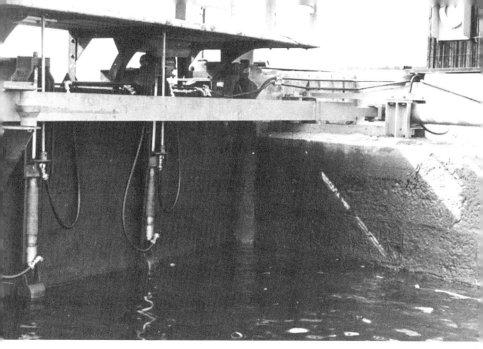

Hydraulic gear for operating lock-gates and sluices.

long; 20 ft yachts paid £7, and 20–40 ft yachts £9 for the through passage. Charges for cargo vessels were actually reduced to a point which would almost always make the through journey cheaper than going by the Pentland Firth, in order to try to win commercial traffic back to the Canal. Given three fair-sized cargo boats a day, the Canal would pay its way but in 1972 there was no sign of them. Thus, neither local industry, not forestry nor commercial traffic was likely to make a significant contribution to the prosperity of the Canal.

What was to be its future, looking ahead from 1972? There is no doubt that the Canal is attractive to visitors who discover it. When two boats pass through the locks at Fort Augustus on a summer day, the lock-side is alive with spectators because something is happening. Cameras click and the scene is recorded: lock-keepers have to be photogenic figures. More ocean-going yachts and motor cruisers were being attracted to the Canal each year, nearly two hundred from sea to sea in 1971 besides almost another three hundred partial passages by pleasure craft. Some yachts from England were laid up in winter in the giant Muirtown Basin because the charges were reasonable. Three separate firms began to hire out motor cruisers at the east end of the Canal and one started boat-building at Burnfoot at the top of Muirtown locks. These developments marked the beginning of a boom in holidays afloat on the Canal.

Since *Scot II*, the Canal tug, was converted to take passengers in 1960, more and more visitors have taken the opportunity to enjoy short trips on the Canal. Making three trips a day in summer, she often carried a full complement of sixty-five passengers. The level of enthusiasm for this service was much higher than it was for a similar service by the *Princess Louise* in the 1930s. This suggests that tourists might welcome the opportunity of longer cruises, such as an excursion to Fort Augustus by steamer and return by bus, or full-day cruises between Banavie and Muirtown, which *Gondolier* and *Glengarry* used to provide. Such day cruises might be very popular if fitted into bus tours of the Highlands and could be operated by two boats, one between Banavie and Fort Augustus, the other between Fort Augustus and Inverness, with a minimum of lockage if passengers changed boats at Fort Augustus. It is well known that it was buses and cars in the 1930s which drove the steamers out of business but the situation has become quite different. The density of road traffic is taking much of the pleasure away from touring by car and bus, and a trip by boat from sea to sea in a day would be a change and a relaxation for visitors and a chance to appreciate the Great Glen and the Canal.

Another proposal, made by members of Glasgow School of Art in a report on Lochaber for the Scottish Tourist Board, would make the Canal itself the centre of interest. This was the provision of a 'boat trail' from Banavie to Gairlochy, similar to a nature trail or forest walk. There would be an information centre at Banavie, explaining the construction, purpose and functions of the Canal, and an opportunity to sail to Gairlochy, stopping at places of interest along the reach. The west end contains the greatest variety of structures on the Canal: Neptune's Staircase, the aqueducts, Moy Burn inlet and the only original cast-iron accommodation bridge. Mucomer Bridge and the new course for the River Lochy out of the Loch are not far away. The proposal was worth serious consideration because it would make the main feature of the district meaningful to all who came to explore it.

If it is true that some millions of people spend part of their leisure 'messing about in boats', it would appear that too few of them knew about the attractions of the waterway through the Great Glen. The

Scot II in action breaking ice.

Canal and the Lochs are not, however, an empty equivalent of the Norfolk Broads. There are few sheltered bays in the long narrow Lochs and the wind can have a funnel effect because of the steep hills on each side. Conditions on Loch Ness and Loch Lochy can sometimes be as boisterous as at sea, but for the experienced and well-equipped amateur the full length of the Canal is available and almost empty.

Since about twenty miles of the Canal is artificial waterway, there is scope for other kinds of leisure activity, from cruising to canoeing, on different parts of it. Consultation with various interested individuals and bodies should lead to the preparation of a comprehensive plan for extended recreational use of the Canal, complementary to its commercial function. It should decide which activities are suitable and safe on each part of the Canal and provide for controlled development of the waterside facilities required. The vast amount of available water space on the Canal needs to be considered in the light of the leisure revolution when larger numbers of people than ever are seeking

Gathering of Canal workers for the presentation of 40 years' service badges to William Gunn, Lock-keeper, and Cyril Wall, Collector of Tolls. R. Brian Davenport, Canal Manager and Engineer, is on the extreme left.

something to do in the Highlands. It is also important to realise that defects for which the Canal has been criticised in the past – the size and number of locks, and the failure to attract industry – have suddenly become assets in the eyes of tourists. Passing through locks could be the highlight of a Canal holiday against a backcloth of natural beauty.

Canal-users need more information about the attractions of places on shore, such as sites of natural or historic interest, forest walks and sporting events. So far, the facilities in the Great Glen have been aimed at the needs of the greatest number of visitors, who arrive by car or bus. Equally, the Canal deserves to be drawn to the attention of land-based visitors. While lay-bys offer splendid views along the Lochs, it is possible to travel sixty miles on the main road between Fort William and Inverness and not realise that the Canal exists, unless a boat in the locks at Fort Augustus catches the eye. When well-tramped walking

Opposite: An unusual monster on Loch Ness, the Hovercraft built by Denny of Dumbarton, on its way to the Thames (1963).

routes, such as that along Hadrian's Wall, become over-crowded and enthusiasm for old canals spreads to the Highlands, the Canal banks could become the haunt of those who wish to tread the paths where horses once pulled sailing ships against the wind; or to see the inland locks, or to stand on top of aqueducts, watching mountain streams tumble under the Canal. The route is obvious enough but information boards could add interest to the walk.

It is a paradox of our time that only when services like barge-canals, steam trains and railways are threatened with closure or extinction do people begin to appreciate them and develop a sentimental attachment to them. Such concern sometimes comes too late but always deserves to be admired. It may stem from uncertainty about the modern world they are living in, or from the boredom or worry of work and the need for an outside interest, as well as from a positive regard for the achievements of early industrial pioneers. Interest in industrial archaeology is of fairly recent growth. It has been much concerned with industrial sites which have been abandoned. New Lanark cotton mills, with their associations with David Dale and Robert Owen, are a good example of this. Elsewhere, old machinery is overhauled and put to work again and is of great interest to tourists. The Caledonian Canal has attracted little attention from this point of view. It is almost as old as New Lanark and is still in good working order. In its field it was as much a pioneer project as New Lanark – it was Britain's first ship canal.

That the Canal was still operating in 1972 is due to several quirks of fate: public ownership since its construction, lack of rail competition until 1914, and its remoteness from other canals. Being under no threat of closure, it did not need a society to preserve it, although it was known and enjoyed by only a few – the fishermen and sailors who use it, people who go for a walk along its banks, and children brought up near it. As an engineering achievement it has been under-appreciated and as an amenity under-utilised. It deserves to be publicised for what it is – a working bit of industrial history which still serves a commercial and recreational purpose in an open unspoilt landscape.

15

JOURNEY THROUGH THE CANAL IN 1972

Clachnaharry to Muirtown Basin

Most passages through the Canal are made by fishing boats. As their custom is to leave their home ports about midnight, boats from nearby are often at the sea-lock at Clachnaharry before dawn, while those from Peterhead can be expected about nine o'clock, usually on a Monday morning. Entry to either end of the Canal depends on the state of the tide. During spring tides when the range between high and low water is seventeen to eighteen feet, boats can be taken into

Keepers of the sea-lock, Jack Cunningham (right) and Tom Garrity, winners of the award for best kept lock on British Waterways in 1969.

the Canal from four hours before high water until four hours after it. In other words, they can enter or leave the Canal during a period of eight hours out of every twelve. The locks are operated between 8 a.m. and 5 p.m. on weekdays all the year round with provision for lockings 'out of hours' on payment of a late locking fee. On Sundays in 1972 the Canal is normally closed but ships may enter or depart through a sea-lock if arrangements have been made in advance.

When the level of the water in the sea-lock becomes equal to the level of the sea by raising the sluice-gates, the lower lock-gates can be opened. Vessels sail slowly into the lock and tie up to the bollards on the lock-side. It is possible for two, three or even four boats, depending on their size, to pass the lock at once. When they have all made fast, the sea-gate is closed. To raise the boats to the level of the reach above, the sluices in the upper gates are opened and water pours through until the levels of the lock and the reach are equal. The gates open and the boats proceed one by one out of the lock into Clachnaharry Reach. They are leaving the sea-lock which was Matthew Davidson's great achievement in 1812 and which won new recognition in 1969 as the best kept-lock on the whole of British Waterways.

Muirtown Basin in the 1970s.

The memory of Thomas Telford remains fresh among the people of Clachnaharry almost as if his last inspection had been made the day before yesterday. His reputation for generosity and his concern for human life are recalled in a story still told locally about the building of this lock. Two children were playing near where the men were working and one of them, a little girl, fell into the sea. She could not swim and immediately a labourer jumped in and rescued her. Telford who was nearby saw what had happened and rushed up to congratulate the man on his prompt action. Then fumbling in his pocket Telford pulled out a beaverskin purse. He threw it to the man without opening it and walked away. It was found to contain £11 in gold coins.

At the end of the reach the swing bridge carrying the railway across the Canal and operated from the adjoining signal box has to be opened to let the boats pass through to the Works Lock. On the right are the workshops, the repair centre for the entire Canal, which gave the lock its name. The carpenters' shop and smithy were built at the time of the Crimean War on the site of the original wooden workshops. The lock also serves as the gauge lock: single locks being ten feet shorter than those in the flights, a ship able to pass this lock can sail through the whole Canal. Telford reckoned that the masonry here was the best on the Canal. In 1971 its maintenance by lock-keeper Willie Gunn won the award for the most improved lock on British Waterways.

Muirtown Basin to Loch Ness

On the right of the entrance to Muirtown Basin is the slipway where new lock-gates are prepared. Beyond it pleasure craft are moored alongside vessels which are used on repair and maintenance work on the Canal. The Basin is big, but comparatively empty, a reminder of the unfulfilled intention that it would become a second harbour for Inverness. A distillery stands where store-houses might have been. Traffic on the A9 stops, the bridge opens and the boats pass through to Muirtown Locks which will raise them thirty-two feet through four locks. The lock-gates look like a series of broad inverted Vs as they are held closed against the underwater sills by the pressure of water above them. When pressure on each side of the lowest gates is equal, the gates open and the boats move in. The gates close and water from the lock above pours in through the sluices until the level in the lowest

Left. Old bollard at Corpach. *Centre.* Revolving capstan for winching sailing ships from the jetty to the sea-lock at Clachnaharry. *Right.* Ship's bell on *Scot II.*

lock rises eight feet to the level in the second lock. The second gates open and the boats proceed, having taken the first step up the flight of Muirtown Locks.

Above the Locks, *Scot II* may be moored on the right opposite the quay constructed in 1876 and used by *Gondolier* and *Glengarry* in the heyday of passenger trips on the Canal. From the turning place for steamers near Burnfoot, made during the reconstruction in the 1840s, the steamers used to reverse back to the quay. Today, there is a new boat-building works at Burnfoot. Above on the right is Craig Phadrig, where the well-known vitrified fort built in the fourth century BC, was re-excavated in 1971. The story that Telford was responsible for an earlier excavation has recently been confirmed by the discovery of a map, drawn by Andrew May in 1812, which shows the location of ten trial pits he had dug down to bedrock in the fort. Farther on to the left is Tomnahurich, a hill like an upturned boat, claimed to be the most beautiful cemetery in the world. The boats approach Tomnahurich Bridge at a speed limited to 6 mph. The bridge-keeper stops the traffic and opens the bridge to let them through towards the sharp right-hand bend round Torvean Hill. Here the Canal is very close to the Hill and high above the River Ness on the left. On through Dunaincroy, over the patch of woollen cloth and clay laid by James Davidson in 1825 to try to keep the Canal watertight, they arrive at Dochgarroch regulating lock, which serves to accommodate the difference in level between Muirtown Reach and Lochs Dochfour and Ness. The level of both Lochs was raised by a weir in the 1840s and the overflow from the weir becomes the River Ness. Between Loch Dochfour and Loch Ness where the channel had to be widened and deepened by dredging, the boats pass Bona, the site of an early and important ferry. Bona lighthouse formerly had a light produced by paraffin or carbide in the keeper's bedroom but changed to electricity outside and is automatically controlled. The lighthouse is now a private dwelling.

In Loch Ness the fishing boats cast aside inhibitions about speed limit with twenty miles of open water ahead. The Loch is long and very deep between steep hills. Here and there they pass villages, Dores and Foyers to the east, Abriachan, Drumnadrochit and Invermoriston to the west. Urquhart Castle is a well-known landmark to sailors and south of it Meal Fuarvonie stands higher than the surrounding hills. The view is of water, coniferous forest, rock and changing sky.

Fort Augustus to Loch Oich

Near Fort Augustus, the former Benedictine Abbey dominates the left of the entrance to the Canal, with a 'pepperpot' lighthouse beside it. On the other side are the remains of the grand but short-lived railway to Spean Bridge. The turn-bridge opened, the boats pass through and enter the lowest lock, which was the most difficult to build on the entire Canal. On both sides are Canal houses, occupied by lock-keepers, retired lock-keepers, and the crew of *Scot II* as well in 1972. The old workshops on the east side became a museum for a time and behind them are the old saw-pit and one engine-house, since the others were demolished in 1886. Old capstans and a windlass are reminders here, as at other locks, of old ways of operating the locks. Raised forty feet through five locks the boats are now above the level of the houses in the village. In 1882 the houses here were in danger when the steamer *Rockabill* smashed a lock-gate. The steamer was swept back by the water above it but the lock-keepers managed to catch it between the jaws of the lower lock-gates. Their prompt action prevented further damage and saved the village from being flooded.

Above the Locks, the old railway swivel bridge has been removed. The fishing boats enter the wide turning-place where *Scot II* spent the winter in the 1970s, convenient for maintenance and icebreaking

In Corpach basin, getting ready to go fishing.

duties on the inland reaches of the Canal. They are now half-way
through the Canal but they have to pass two more locks, Kytra and
Cullochy, before they reach the Canal summit at Loch Oich. All
the way the Canal and the River Oich are close to each other and
in places, for example north-east of Kytra, the Canal is in the old
river bed. These locks are lonely inland places where the tradition of
lock-keeping runs in families, the Bains at Kytra and the Kennedys at
Cullochy. Andrew Bain, who retired to Fort Augustus, and his father
before him, operated Kytra Lock for the whole of their working lives.
The Kennedys have been at Cullochy since 1919 and Dan, the present
lock-keeper, has worked there for forty-seven years. His brother Alec
is one of the lock-keepers at Fort Augustus with nearly thirty years
service on the Canal. A regular job on the lock, a free house and a
bit of land as grazing for a cow – these are the ingredients of a simple,
satisfying way of life. The fishing boats bring news of the sea and the
fishing as they pass on their way to the loneliest and loveliest of the
Canal Lochs, Loch Oich. The islands in it are signs of its shallowness
and of the amount of dredging that was necessary to make it navigable

through the centre. The ruins of Invergarry Castle and probably the spirit of Alastair MacDonell of Glengarry as well still scowl at passing ships.

Laggan to Gairlochy.

Out of the Loch, the boats swing into Laggan Reach, the deep summit cutting between an avenue of trees on its broad artificial banks. A stream on the left flows into the Canal over an artificial waterfall; others enter through culverts under the banks. At the double lock at Laggan the V-shape of the lock-gates is now pointing towards the fishing boats, which are about to take their first step down from the summit. Leaving the lower lock they enter Loch Lochy, the last of the freshwater Lochs in the Great Glen. The Loch is ten miles long, its shores regular, the hillsides steep and wooded on the lower slopes. Towards the south-west the Loch widens towards the Bay of Bunarkaig in the west and opens up on the other side to give splendid views of Ben Nevis and the surrounding hills.

A little lighthouse marks the end of the Loch and the approach to the locks at Gairlochy. The upper lock, constructed by Jackson & Bean in 1844, is the only additional lock built on the Canal since the time of Jessop and Telford. The original block of lock-keepers' houses still stands with the stables for tracking horses behind them. The retired lock-keeper at Gairlochy, John MacLaren, who has been succeeded by his son, still recalls his busiest day when fifty-six drifters were passed through in sixteen lockings. Mrs MacLaren is the daughter of Alexander MacIntosh, who was a carpenter at Clachnaharry before the First World War, and has connections with Dochgarroch, where one of the lock-keepers in 1826 was Donald MacIntosh, a carpenter. This is one of the families on the Canal who are descended from the original Canal builders and have a tradition of continual service covering over a hundred and fifty years.

Gairlochy to Corpach

After seeing the boats through the locks, the lock-keeper in 1972 would set off by motor-cycle to open Moy Bridge, a mile and a quarter away. The last of the original accommodation bridges on the Canal, it gives access to the fields of Moy farm. The lock-keeper opens the half on the south side first by turning a capstan by hand, and then he rows across in a punt to open the other half. When the boats pass he closes the bridge. The cottage here, built as the bridge-keeper's house, has windows in each gable to give a view in both directions along the Canal. Upstairs in an attic bedroom there is a small window, only nine inches square, through which the bridge-keeper could keep an eye on the Canal even from his bed. Nearby, the Moy Burn flows directly into the Canal because it proved too difficult to build an aqueduct over it, but farther on the boats sail over the series of aqueducts across the River Loy and mountain streams to the top of Banavie Locks.

Controlled by two lock-keepers Goodwin and Macpherson in 1972, the passage down Neptune's Staircase takes less time since mechanisation than it did in the old days when twelve men were employed. The benefits of hydraulic operation are most evident here, where there are eight locks to pass. In 1870 news of massive herring

The Pulp Mill at Corpach in 1971 with the Canal entrance beyond on the left, Ben Nevis and Fort William on the right.

Left. Duncan Cameron, Master of *Scot II* at the wheel. *Right.* Alec Kennedy, Lock-keeper at Fort Augustus in 1971.

shoals off Barra and in the Minch brought an almost continuous queue of five hundred and twelve fishing boats through from the east coast. Opening and closing nine pairs of lock-gates by rotating capstans by hand to pass so many boats through must have been the most exhausting experience in the lives of the lock-keepers then at Banavie.

The boats are lowered sixty-four feet in the space of five hundred yards and pass Banavie Station, through bridges for road and rail which connect the Canal with Fort William. Crossing Corpach Moss to the double lock above the Basin, the boats come under the charge of lock-keepers Burnell and McLachlan who see them through the Basin and out at the sea-lock. The Basin handled trade for the Pulp Mill in 1972 as well as being a favourite haven for boats fishing on the west coast. The 'pepperpot' lighthouse was built in 1913 but the old engine-house, a feature at Corpach since the construction of the sea-lock, was demolished in 1968. It had been converted into houses for lock-keepers, and store-rooms, one of which became a little salmon-curing factory for a time, and another a bobbin factory using birch wood from the slopes above Loch Eil. The head lock-keeper in 1972, formerly a railwayman, was a railway enthusiast who settled for the timelessness of the Highland Canal.

At each lock along the Canal, the keepers log the name of each vessel and the time when it locks. It is at Corpach that payment is made by vessels proceeding in either direction in 1972. As far as ship-owners and the Collector of Tolls are concerned, the cost of a single passage is reckoned in £s and pence. The real costs, however, include a share of the expense of operating and maintaining the Canal by fifty men and, in addition, the displacement of nearly sixty thousand cubic yards of water. For crews making for the fishing grounds, the value of the Canal is measured by the amount of time they save by passing through it. Depending on the amount of traffic, their journey through the Canal will have taken about twelve hours.

16

AN OLD CANAL RESTORED

23–24 October 1972 marked the 150th Anniversary of the ceremonial voyage which opened the Canal. Whereas its centenary had been allowed to pass almost without notice in 1922, a hundred and fifty was considered a fair old age for a working canal and the arrangements the British Waterways Board and the Institution of Civil Engineers made jointly to honour the occasion owed not a little to how things had been done in 1822. A water-borne progress by landed proprietors would not have been appropriate, but giving the public the opportunity to inspect the whole length of the Canal by boat and bus was, and it proved very popular at £3 per head.

The party who sailed on *Scot II* first along the Canal and through Loch Ness found themselves being welcomed into Fort Augustus like the very important people they were that day by music played by the brass and pipe band of HMS *Caledonia*. The bus party, who explored the west end first, encountered ample evidence of the Canal's usefulness to fishermen. They saw the bridge open at Aberchalder as two fishing boats passed in front of them into Loch Oich; at Corpach they found fishing boats sheltering in the Basin over the weekend and, as they stopped to inspect the Loy Aqueduct, more fishing boats were passing just above their heads. On their return on board *Scot II* along the unlit Canal to Muirtown, darkness was falling but they were safe enough. Duncan Cameron, the master, had been sailing the Canal since his days on the *Gondolier* and, a voice assured the passengers, 'He navigates this Canal by instinct.'

The BWB gave an anniversary dinner in the Caledonian Hotel, Inverness, for representatives of local authorities, public bodies and businesses connected with the Canal. The Board's Chairman, Sir Frank Price, took the opportunity to attack Government proposals to break up his Board, hand over English canals to regional water

authorities and divide up the five Scottish canals among the local authorities in whose territories they happened to be. Dr Angus Fulton, a past President of the Institution of Civil Engineers, said some wise things about Thomas Telford who, besides completing this Canal, was the first President of his own Institution.

Mr Brian Davenport, the Engineer responsible for the Canal, delivered a lecture about it to a large audience in Inverness Town House. Many other lectures that winter carried the story of the Canal past and present to people in different parts of Scotland. One of the most interesting discussions took place at Fort Augustus where an audience ranging from pensioners to Cub Scouts heard lock-keepers and one of the crew of *Scot II* helping to answer questions from their own experience. The effectiveness of talk and pictures is always difficult to gauge but information was spread and seemed to give pleasure, and the Canal's name began to feature more prominently in tourist maps of the area.

Canal Studies

This book was first published in October 1972 on the 150th Anniversary of the Canal opening.

In *William Jessop, Engineer*, published in 1979, Charles Hadfield and A.W. Skempton, by careful study of the sources, have added to our knowledge of how important Jessop's contribution was to the construction of the Caledonian Canal during the first ten years from 1803. It was on the basis of his estimate in 1803 of what the total cost would be, for example, that Parliament passed the main Caledonian Canal Act in June 1804; and when Telford suggested how valuable 'the assistance and advice of Mr Jessop' would be on his next visit to the Canal, the Commissioners decided to give Jessop a continuing role and made him, in effect, consulting engineer. Normally visiting the Canal once a year, Jessop received payment at the rate of 5 guineas a day, compared with Telford who came to the Highlands twice a year and whose rate was always 3 guineas a day. Theirs was a professional partnership which allowed this distant Highland Canal to benefit

Opposite: The Strone sluices, from a French book in 1828: note the young coniferous trees, the sailing ship and perhaps the only known picture of a tracker.

from the canal-engineering experience and skills they had acquired in the south.

The book established also that it was Jessop who designed the first dredger for the Canal, and who chose the same type of bridge for the Canal as for the West India Docks where he was Engineer, while Telford acknowledged that building the bridges of cast-iron instead of wood was a decision they made together.

Thomas Telford: Engineer was published in 1980, edited by Alastair Penfold, who wrote the paper in it on 'Managerial Organization on the Caledonian Canal'. He demonstrated that Thomas Telford appointed the Superintendents, Matthew Davidson and John Telford himself, and shed some light on the earlier career of John Telford, who had worked with Thomas Telford on the Ellesmere Canal and become his personal assistant and draughtsman, making the plans for Bridgnorth Church (1792), for example. Jessop gave these appointments and those of the masonry contractors the nod of approval and left lower-level appointments to Thomas Telford's discretion. Superintendents appear to have chosen the general contractors, having been advised that their job of managing the work would be made simpler if they employed only a small number of firms of contractors.

Thomas Telford's Temptation by Charles Hadfield appeared in 1993, with the subtitle *Telford and William Jessop's Reputation*. In his previous book on Jessop with Professor Skempton, *William Jessop, Engineer*, they had constructed a detailed and positive account of his career as a civil engineer, in spite of the absence of his personal papers, and had succeeded in drawing attention to Jessop's wide-ranging achievements, but particularly his canals, docks and harbours. In this book, his aim was very different. He set out to reveal a subtle campaign by Telford to diminish the part Jessop had played in the projects in which they were both involved, either by omitting or forgetting to mention Jessop's contribution, and thereby to win most of the credit for himself.

He marshalled a mass of detailed evidence to show that this was not simply the selective memory of a self-important, tired and deaf old man at work, as he tried, and failed, in the 1830s to breathe some life into his *Life of Thomas Telford, civil engineer, written by himself.* He cited the influence of friends who received all their information

Opposite: Thomas Telford when he was 72 by George Patten, a little known portrait in Glasgow Art Gallery.

from Telford, such as Robert Southey, the poet laureate, who first met Telford in 1819 when he toured the Highland works with him and wrote enthusiastically above the wonders he saw and Telford's key role in creating them:

> Telford's is a happy life: everywhere making roads, building bridges, forming canals and creating harbours – works of sure, solid permanent utility.

Another influential friend was John Rickman, Secretary of both Commissions, for the Caledonian Canal and for Highland Roads and Bridges who, in preparing the *Life* for publication after Telford's death, did his best always to make sure that it did honour to Telford's reputation.

Hadfield accused Telford of amending the record of the Pontcysyllte Aqueduct within months of its opening in 1805 to secure as much of the favourable comment it stimulated for himself, and reinforcing his long personal association with it by using it as the background to the portrait Samuel Lane painted of him, which Telford presented to the Institution of Civil Engineers in 1826. Hadfield's evidence is persuasive and it looks as if I was too generous to Telford when I suggested on page 27 that he would have wished Jessop's contribution at Poncysyllte to be fully acknowledged. The fairest conclusion must now be that the Aqueduct was their joint achievement along with the others named on that page. As consulting engineer Jessop's role was predominant in the earlier years, taking the strategic decisions on its line, its dimensions, the materials to be used and how it was to be built, with Telford his conscientious agent on the spot taking on responsibility for bringing the project to completion and staying on as Engineer of the Ellesmere Canal for the rest of his life.

While they were both engaged on the Caledonian Canal, Telford began his long association with the Göta Canal in Sweden, a sea-to-sea canal through lakes like the Caledonian, to link Göteborg in the west with Mem on the Baltic. Telford who visited Sweden in 1808 to survey the route and recommend the project and later supplied plans and key personnel for its construction, seems to have given the Swedes the impression then and in twenty years of correspondence with its promoter, Count von Platen, that he was the only engineer on the Caledonian Canal, never once mentioning the name of William

Jessop. Perhaps the wheel of reputation has turned against Telford now, even if unfairly. The guidebook to the Göta Canal in English in 1991 contains no reference whatever to his name, all the credit for the project going to von Platen.

After the publication of Telford's *Life* Samuel Hughes was, as Charles Hadfield reminds us, the first to point out in 1844 that William Jessop's contribution to the Caledonian Canal was in danger of being overlooked, insisting:

> Mr Jessop did originally design the works, and settle the principal points, yet in the execution of these works . . . these two engineers are entitled to share alike in all the merits and defects which belong to them. It will add weight to this opinion to state that it coincides with that of Mr George May, of Inverness, the present talented engineer of the Caledonian Canal Commissioners.

Allocating praise or blame is never easy. Hadfield, for example, emphasises that 'the Banavie staircase had been designed and built (except for the lock-gates) in Jessop's time'; meaning that most of the credit for it must go to him whereas we have seen on page 114 that George May, mentioned above, had condemned the masonry of the whole structure as 'execrable', expressing his surprise that Telford (not Jessop) had been taken in! This is a reminder that Telford, making his second inspection tour of every year on his own must have been inspecting all the works, just as Jessop would have done, as well as checking the accounts, on behalf of the Canal Commissioners. Jessop and Telford were a team for ten years and Hughes's conclusion that they 'are entitled to share alike' the credit or blame for the quality of the structures is probably as far as it is wise to go. By visiting the Canal twice a year until 1814, however, Telford must have been the better known to the workers along it and, being in sole charge in the years that followed, it is not surprising that his is the name associated still in the popular mind with the Caledonian Canal.

In 1990 a request for permission to translate part of this book *The Caledonian Canal* into Japanese came as a complete surprise. It came from Mr A. Nagai whose firm had expanded into Telford New Town in Shropshire, which took its name from Thomas Telford. His book, *Thomas Telford and the Caledonian Canal – The Human and Social Aspects* (1991), contains his own introduction, Telford's early life translated from Samuel Smiles's *Lives of the Engineers*, Telford's plan for the Caledonian Canal from Sir Alexander Gibb's *The Story of Telford*, and the construction of the Canal, Chapters 4 to 9 of this book.

Maintenance and Repair

In 1971 the Canal Engineer moved south to become Engineer Scotland in Glasgow, where engineering services for all the Scottish canals were centralised, and was not replaced. For the Caledonian this was the end of an era, because there had always been an engineer on the spot, living in the house beside the Canal Office, since Matthew Davidson arrived to start building the Canal back in 1804.

The Canal closed for six weeks for an overhaul in the spring of 1973. This was a scheduled closure, nine years after the previous one. *Ocean Mist* at the top of Banavie Locks and other boats on the Canal were moved to the lochs but when all the water was run out of the canalised sections, is it true that there was a telephone box standing upright in the middle of one of the locks at Muirtown? Lock-keepers helped the repair squad in cutting back branches and bushes and filling the worst of the holes in the banks with stones which they covered with wire mesh. To try to fill cavities behind lock walls and under sills where water was always pouring through, concrete was pumped in under pressure and the joints between the stones were re-pointed. Divers carried out the repairs on the locks it was not possible to drain.

The repair men would have liked a longer closure and enough money to do a good job, whereas sailors, fishermen, hire fleet operators and, indeed, canal managers want to see canals open and operating. The timing of closures and the length of time they last are always compromises taking account of a variety of conflicting interests, while the amount of money available is decided ultimately by Government which has many priorities which are apparently higher than filling up holes in an old Highland canal.

Events were to prove that the repairs done in 1973 were not enough. Twice in 1975 the Canal had to close for repairs and 1976 was a year of two crises. On 24 May when Douglas Forman of Aberdeen was in the middle of Muirtown Locks in a new cabin cruiser he had brought up from the Bristol Channel, the gates behind him were suddenly upended and disappeared under the surface. Hearing the crash he looked round, 'It seemed the sky had turned into a wave of green water. There was no time to feel fear. The next thing I remember for certain was the lock-keeper reaching down and grabbing my arm.' While he was being rescued, water poured unhindered into Canal Road and gardens on the other side. The repair took six weeks.

Then, exactly a century after the accident at Laggan Locks when the steamer *Staffa* smashed the gates and closed the Canal, Parliament heard that the Canal was blocked again at Laggan Locks by a major wall collapse. Because the British Waterways Board had no money of its own to cope with such an emergency and the Government stood firm so long, the closure lasted for ten months.

These closures gave point to a lesson which was only slowly being learned: that old canals, like old people, need more care. The Report to the Department of the Environment in 1975 by Peter Fraenkel and Partners put this view about the canals of Britain as a whole. It demonstrated that the British Waterways Board had not done all that it should to maintain the canals in good order because it had too little money and that a backlog of maintenance existed, requiring close on £40m. of extra expenditure at March 1974 prices, about £60m. by the time the Report was at last published in November 1977. For economy's sake and the safety of the canal system, the need for action was urgent. The Caledonian Canal was singled out for especially extensive repairs, likely to cost £911,000 at 1974 prices. The lock chambers in particular required so much grouting, filling with concrete under pressure to make them stable, that 36 per cent of the expenditure to repair the lock structures on all the Commercial Waterways was recommended for the Caledonian. It is interesting to note that it was a lock wall collapse at Laggan Locks that closed the Canal in 1976 and that at that moment the Fraenkel Report was, if not on the desk of the Secretary of State for the Environment, then certainly on a shelf somewhere in his Department, and it may have played some part in the decision, after four months' delay, to release funds for repairs to go ahead.

The work was done by Edmund Nuttall Ltd who first erected coffer-dams to hold back the summit water above them and Loch Lochy below, and then drained the locks. For the first time since they were built over a hundred and fifty years before, men could walk

Laggan upper lock empty, protected by the coffer-dam, grouting and power-pointing completed.

about freely on the bed of the locks. The main reconstruction work was on the recesses behind the lowest gates. After driving sheet piles to secure the ground behind, they dismantled the masonry, carefully numbering each facing stone in order to put it back again later in exactly the same position. This was to preserve the character of these locks which are 'listed' as a building of special architectural and historic interest. The original wooden sills were scrapped and replaced in concrete. Over all the walls, grouting specialists pumped plasticised mortar under pressure into the joints between the stones to make them all secure. Or almost all: when last seen a persistent jet of water kept bubbling up above the floor and, who knows, it bubbles still. Laggan Locks empty that season attracted many visitors who came by hired cruiser and by road. The Canal Centre, a tea-room with an exhibition about the Canal, which had opened in the little shop beside the Locks, proved a popular place, not least with hungry workers on this lonely site. On 29 August 1977, too close to the season's end for cruisers but at an opportune time for fishing boats, the Canal was open again.

Other weaknesses Fraenkel highlighted on the Caledonian Canal were the state of the lock-gates and the canal banks. Some of the gates had been in use since before the First World War and had become twisted over the years, especially through being operated by being pushed and pulled at the top. A programme had already begun to install steel gates fitted with buoyancy tanks, which open and close more easily and will not bend. By the end of 1977 twenty new gates had been installed, by the end of 1981 forty-three, and all the gates had been replaced by 1989 with the exception of the wooden gates on the lowest lock at Fort Augustus which survived until 1992. As the lock-gate replacement programme progressed, *Gatelifter III* pictured on page 95, which had formerly lifted gates out and in from the water gave way to a huge hired crane working from the side of the lock. Old gates weighed about 20 tonnes whereas the new gates, 29 feet high and 24 feet wide, were lighter.

Where the wash from passing vessels had dislodged much of the stone pitching, the banks were in danger, and the use of wire baskets, called gabions, filled with fairly big stones, is the new way of protecting them. Laggan cutting with its steep sides was one of the worst stretches, gouged by landslips and falling trees, and there the trees were felled and extracted overhead across the Canal and the worst-holed banks were repaired. A speed limit of 5 knots was enforced on vessels by refusing to allow them through the next lock before their predicted arrival time. As a vessel's size is also a factor in the damage it may cause, the absence of big commercial craft from the Canal now may be a blessing in disguise.

Identification of the Canal's defects in a national report along with estimates of the costs of rectifying them could do nothing but good, provided sufficient finance kept flowing to maintain the momentum. There was real hope that it could be brought up to a better condition than for years past. Where stonework is in constant contact with water and peat, of course, there is always some danger, and vessels using the Canal can cause accidents too, like the barge which hit Moy Bridge in January 1981.

The 1980s have indeed witnessed a sustained campaign to restore the lock chambers, especially those in the centre of the Canal which had been the last to be built. Closures were planned for the winter months when the Canal had least traffic. Work started in October 1983 on the lowest three locks at Fort Augustus, which are below even the lowest water level of Loch Ness and had not been repaired in dry conditions for a hundred and forty years. As water was the enemy during the construction of these locks and again during their reconstruction in the 1840s, keeping the site dry was also a challenge for the engineers this time. A coffer-dam held back the water of Loch Ness but all the water flowing into the Canal above the locks from Kytra Lock and the catchment to the south was diverted through a pipe through the Canal bank into the River Oich. The three lower sills, or steps up to the next lock against which the lock-gates close, were completely replaced in concrete by Morrison Construction Ltd during the worst winter for years. A canal being emptied in a village ought to reveal some secrets but as M. Lansdowne, then an Abbey schoolboy, reported, 'When the Canal was drained, few things of interest were found: some clayware bottles, clay pipes, an 1875 shilling, and many eels.'

Muirtown and Clachnaharry sills were repaired in November 1987 and Kytra Lock received a thorough overhaul by Fairclough (Scotland) Ltd in the first three months of 1988. Not only were the gaps behind the lock walls filled, the walls repointed and the sills replaced in concrete but a special cement wall was built down into the bank to a depth of 13 metres at right angles to the lock walls to stop water seeping through them again.

The need for emergency repairs at Neptune's Staircase at Banavie required the west end of the Canal to be emptied of water in February 1989. As at Fort Augustus in 1983, it was water leaking through deteriorating sills that was causing the damage, making the gates difficult to open and close and putting the walls of the locks in danger. Here too three new concrete sills were constructed, this time by R.J. McLeod, contractors of Dingwall, while others were repaired as far as possible. A pumping operation kept Corpach Basin open all this time to allow coasters to continue to bring in pulp for the paper mill.

At Cullochy, the last of the original locks to be built on the Canal, consulting engineers, Robert H. Cuthbertson and Partners found in

Opposite: Aerial view of Cullochy lock empty as restoration work went on despite the snow and the Canal frozen above it.

Brian Clark's repair team restoring Moy inlet in 1993, the wooden centering they have erected here a good example of the support required during the construction of stone arches.

1988 that every time the level of the water in it rose and fell, water was flowing freely through the lock walls and behind them. The west wall was already bulging and the banks were in danger. Built directly onto rock like Kytra Lock, this lock also lacked an inverted arch, and showed no sign of puddle and clay ever having been used to seal behind its walls. It also had too few buttresses, which keep the lock stable, and none at all behind the gate recesses which, because they are vertical and not curved like the walls of the lock, are the points of greatest weakness. With the lock still full of water, tube à manchette grouting down to solid rock formed a protective layer behind the walls and made it watertight again. When it was emptied, pressure pointing sealed between all the masonry blocks inside the lock, and on the outside the clusters of mini-piles Fairclough (Scotland) constructed as buttresses and anchored to bedrock will keep it stable. The old Cullochy masonry lock has been cradled in concrete and made safe, with new sills and new gates, at a cost of over £900,000 in 1991, nearly as much as the total cost of building the entire Canal up to 1822, such has been the extent of inflation since then. Although so much of their work is underground or under water and therefore unseen,

Robert H. Cuthbertson and Partners' achievement at Cullochy was recognised when the Saltire Civil Engineering Award was presented to them in 1992.

This programme of phased improvement, restoring masonry structures and installing new lock gates, appeared to fulfil the hopes expressed after the publication of the Fraenkel Report. But water flooding in behind the walls of many locks every time the locks were in operation, posed a constant, though unseen, threat. After a lock wall at Fort Augustus crumbled in 1995, it was decided to conduct a radar survey on all the locks. For the first time ever it revealed the true extent of the damage – huge holes in the rubble walls behind the lock walls, due to the lime mortar dissolving, and inside that, great gaps in the clay puddle lining which had been intended to keep the locks watertight.

Clearly the Canal was in a dangerous condition. It must either be closed, which it was estimated would still cost about £1m a year to maintain it, or undergo a wholesale programme of restoration. This would involve pumping into every lock wall clay-based material to replace the clay puddle and then grout to firm up the rubble inner walls.

The need was urgent but the cost, £20m, was far beyond the Canal's normal resources. In April 1996, British Waterways asked the Department of the Environment for £20m of emergency funding, £3m of it to complete the work at Fort Augustus and some £5m for Banavie. Initially the Department of the Environment gave a grant of £2m. The Scottish Office had no direct responsibility for the Canal but, influenced by the arguments that 740 jobs in the Great Glen and over £14m of tourist spending per year depended on the Canal remaining open, it broke new ground by giving two grants, amounting to £4m. During two winter closures from 1995 to 1997 R.J. McLeod of Dingwall contractors completed the restoration of Fort Augustus Locks and then Banavie became their focus of activity.

Splendidly restored between 1996 and 2001 Neptune's Staircase's appearance has also reached a level unequalled before. Its grey granite cope-stones and quoins have reinforced the lock walls and also lightened their look, while having all the lock-gates designed to have water falling continuously over them in a series of waterfalls when the locks are not in use, has turned the whole flight of eight locks into a

Transformer for Foyers power station which came all the way from Manchester by water at Muirtown Locks in 1973.

giant cascade. Floodlit at night, Neptune's Staircase has become the tourist attraction this spectacular industrial monument deserves to be. Harnessing the falling waters here to generate electricity was explored but had to be abandoned.

On an inspection visit in September 1999 John Prescott, deputy prime minister and head of the Department of the Environment, promised that the rest of the money required for the five-year programme would be provided, allowing restoration work to continue on the locks in the east.

The road bridges over the Canal, although operated by Canal workers, are provided by the Ministry of Transport. In 1980–1 they also were renovated, and Laggan and Aberchalder Bridges were raised 60 cm, enough to allow nearly all types of motor cruiser to pass underneath without halting road traffic.

While the restoration work at Banavie was going on, most attention was being focused farther south. This was the massive project to restore, and where necessary reconstruct, the Forth and Clyde and Union Canals and make them navigable again at a cost of £84.5m. Called The Millennium Link, it allows the Forth and Clyde to become a connection between the North Sea and the Atlantic once again, like the Caledonian Canal, and by using the Union Canal for part of the way connects Edinburgh by water again with Glasgow. Forming the vital link between the Forth and Clyde and the Union Canal, which used to be done by passing eleven locks at Camelon, is the Falkirk Wheel, a rotating lift 115 feet high, which is unique and can transfer boats directly from one level to the other. It is remarkable to discover that all the engineering work, designing, constructing and installing the Wheel was done by the Butterley Company in Derbyshire. This is the same company in which William Jessop was a partner and which did so much of the engineering work for the construction of the Caledonian Canal nearly two hundred years ago.

The Forth and Clyde Canal now being open makes it possible to sail from the Firth of Forth to the Firth of Clyde, through the Crinan Canal to Loch Linnhe, and through the Caledonian Canal into the North Sea and back to complete a circular voyage round much of Scotland. The interest aroused by the opening of the Millennium Link in May 2002 has benefited all the Scottish canals.

The Caledonian Canal has no problem over water supply and seldom needs dredging except at three points, Clachnaharry sea-lock, Moy inlet and the Spout at Laggan. The Reservoir Act 1975, however, which regarded Loch Lochy, Loch Oich and Loch Ness as reservoirs for the first time, laid on British Waterways new responsibilities for protecting them against the probable maximum flood. This led to the Gairlochy Flood Relief Scheme, also designed by Robert H. Cuthbertson and Partners, who raised the top gate of Gairlochy Upper Lock by four feet and ran a substantial flood barrier from it along behind the lock-keepers' houses. They also cleaned out the Mucomer cut back to its 1843–6 dimensions, to improve the flow of water out of Loch Lochy in an emergency, and felled all the trees, to prevent a tree ever blocking one of the power station's three 90-feet automatic sluice gates. It will be recalled that floods three feet higher than the lock-gates of the single lock at Gairlochy in 1834 had threatened all the farms between Loch Lochy and the sea with flooding, and had been followed by the deepening of the River Lochy channel and the building of the new upper lock at Gairlochy as flood control measures. These flood prevention works were completed in 1988.

When high tides coincided with days and days of torrential rain in

Restoration of Muirtown Locks, 2004.

February 1989, the River Ness burst its banks and flooded the streets of Inverness, and swept away the railway viaduct. Canal workers were able to help to some degree by diverting water from Loch Ness through the lock sluices down to the sea. At Cullochy the water was two feet higher than the lock-gates at one time and the River Oich, 7 feet higher than usual, flooded houses at Fort Augustus and forced people to leave their homes.

On 25 October 2003 the Canal was closed for the restoration of Muirtown Locks during five winter months. All the craft in the marina in Muirtown Basin were moved up to safe moorings at Dochgarroch and a coffer-dam was constructed above the Locks to hold back the water and allow work to go ahead. The Locks were drained and the water level in the vast Muirtown Basin below them was being lowered ever so slowly to prevent any of the banks collapsing, until the east end of the Canal was open to the sea and subject to the inflow of the tides. In November a wall was being built below the lowest lock in Muirtown flight to protect the works from flooding. Emptying part

Car being removed from Muirtown Basin, 2003.

of the Canal like this always reveals some surprises: in Muirtown top lock three cars were discovered and in the Basin a Nissan Sunny.

The contractors again were R.J. McLeod and the project manager for British Waterways was Paul Colenso. The pattern of the work on these four locks was similar to that already carried out at Fort Augustus and Banavie – consolidation of the walls, repair of the sills and the installation of new lock-gates which cascade like Neptune's Staircase – at a cost of £2m. A lighting scheme has also been installed, which makes it an attraction to the public and also draws attention to the deep water in the locks, a sensible precaution in such a built-up area.

Closure for another winter in 2004–5 saw R.J. McLeod's team, with George McBurnie as British Waterways' project manager, busy upgrading Clachnaharry Works Lock and the Sea-Lock. It also proved possible to restore Dochgarroch Lock in the same season. With these works completed, the structures along the Caledonian Canal have been given a new lease of life.

17

CHANGES IN CANAL TRAFFIC

In 1972 the Caledonian Canal was classified as a Commercial Waterway and most of its customers were commercial. Usually they were fishing boats, and a typical passage was described in Chapter 15. In 1972, 670 fishing boats went right through, compared with nearly 300 yachts and 36 cargo boats. They also found Corpach Basin a good base to fish from and recorded four hundred trips out to sea and back. Between 1975 and 1982 the number using the Canal has declined to about half, partly because of the occasions when the Canal has been closed, but mainly due to the ban on herring fishing in the Minch from 1978. The importance of the Canal to fishermen in the 1970s, however, was emphasised by Gilbert Buchan, speaking for the Scottish Fishermen's Federation in January 1977:

> A great many fishermen regard the Canal as absolutely essential. Being obliged to sail via the Pentland Firth, some crews on the smaller type of vessel are risking life and limb. It is just hopeless there at a spring tide, especially with a north-west wind.

The number of fishing boats passing through the Canal settled at about 95 in 2012 and through passage by a cargo vessel, other than the occasional dredger, has become an unusual event. Most commercial activity tends to be confined to Corpach Basin. In the 1970s the most important cargo was pulp, running at about 25,000 tonnes a year. About 7,500 tonnes of timber used to come in each year for the Pulp Mill, mainly in 80 tonne loads from Mull on the puffers *Eldessa* and *Marsa*, which often carried back building materials. When the Pulp Mill closed in 1980, these timber imports came to a sudden end. With new technology and a new product, however, paper-making continues and 1,100-tonne cargoes of pulp for the Paper Mill come in regularly after a six-day voyage from Spain and now sometimes from Scandinavia and Portugal. Fuel oil goes out regularly as well as coal

Fishing boats at Banavie in 1971.

in bulk for the Island coal clubs while occasional imports of sawn timber and logs from Raasay, and salt from Carrickfergus for Lochaber roads in winter can still be expected. Interestingly the Norwegian MV *Kanutta* introduced a new freight service on the Canal in 2010 and the Great Glen Shipping Company at Corpach is now looking for a bigger boat.

In spite of the decline in commercial traffic, the total number of craft of all kinds passing Tomnahurich Bridge in either direction rose from 2,500 in 1972, to 3,712 in 2012. The number of yachts going right through was also rising gradually (346 in 1980 and 996 in 2012), but the main reason for the staggering increase in boat movements east and west has been the success of hire cruiser holidays on the Canal. A typical Canal journey by 2012 was by a family in a cruiser, between the top of Muirtown Locks and the top of Banavie, taking a week there and back in their own time. Hire cruisers accounted for 1,589 boat movements in the year 2012.

Leisure Use of the Canal

In February 1971 Sir Andrew Gilchrist, the Chairman of the Highlands and Islands Development Board, speaking to the Institute of Marketing in Glasgow, summed up the opportunity his Board saw

on the Caledonian Canal in this sentence: 'There are 2,000 boats on the Norfolk Broads, 200 on the Shannon, and just one on the Caledonian Canal.' The one on the Caledonian belonged to Jim and Elizabeth Hogan who advertised their two-berth cruiser for hire in a Sunday paper the previous summer. They were so encouraged by the response that they bought their first five-berth cruiser. This was the beginning of Caley Cruisers, which became the biggest hire cruiser business with 40 motor cruisers on the Canal. Others who started in 1971 were North Highland Charters and Loch Ness Marine who also built boats at Muirtown. By 1976 six firms were hiring cruisers and, to a lesser extent, yachts on the Canal, including one based in the west at Banavie.

The financial assistance the Highland Board offered in the form of grants and loans, to cover up to 50 per cent of the cost of new boats, moorings and jetties, was invaluable to these new enterprises. One reason for their early success was the decision to have bookings handled nationally by the established cruising holiday firms, Hoseason's and Blake's, who advertised them in the same brochure as holidays on the Broads, the English canals and the Shannon. The Caledonian Canal appealed immediately because it offered the prospect of a challenging cruising holiday in a new area where the waters were uncrowded and the scenery rugged and Highland. Many who came had experience of cruising, usually on the Broads, and over 5 per cent of the bookings came from abroad. By 1976, with seventeen new boats available for hire, canal cruising had become a popular form of marine recreation and a fruitful new branch of the tourist industry. The closure of Laggan Locks affected bookings, but cruising went on in 1977 with seventy craft for hire and over 10,000 people taking this kind of holiday. That year too, the author was encouraged to publish *Getting to know . . . The Caledonian Canal*, a guide to the Canal and places of interest along it for visitors, and two years later Caley Cruisers published a map in colour, *The Caledonian Canal including Loch Ness*, specially for hire boat users.

The popularity of hire cruiser holidays here took nearly everyone by surprise. Their growth outstripped the provision of all the waterside facilities required, and put the Canal's locks and bridges under strain through being almost constantly in use. Early on, eccentric journeys

Opposite: Cruisers under instruction, heading for Loch Ness.

were sometimes undertaken: one crew took a cruiser down Neptune's Staircase 'because it was there', not thinking that it took a lock-keeper half a morning to see them down; while another passed down five locks at Fort Augustus just to collect groceries at Leslie's shop!

When Jacobite Cruises of Inverness began in 1975 with three water-buses and a cruise vessel, ranging in capacity from 50 to almost 100 passengers, tourists had many more opportunities to go on day trips on the Canal, Loch Ness and Loch Linnhe. One water-bus operated between Banavie and Gairlochy that season, and *Neptune's Lady* again in 1988 but not enough customers were attracted. Now called Jacobite Experience Loch Ness it has been operating on Canal and Loch for 37 years and offers cruises on Loch Ness in winter.

The suggestion to provide trips daily between Inverness and Banavie, like the *Gondolier* used to do, has not been taken up, but *Lord of the Glens*, a converted Dutch passenger cruiser with twenty-six twin cabins, offers luxury cruises from Inverness through the Canal along with the *Scottish Highlander* and *Fingal of Caledonia*, which offers activity holidays. They all arrange visits to places of interest on shore along the Great Glen.

On-shore chalets have been built, and on Loch Oich there is The Great Glen Water Park, where canoeing, sail-boarding and water-skiing are among the activities on offer. A comprehensive plan for leisure activities in the Great Glen is now in being under the Great Glen Ways Initiative.

Quite early, the Waterways Board monitored the rate of development, and in 1974 predicted that capacity point for hire cruisers, 60–80 vessels, would come in three years. The prediction was correct about the numbers, seventy in 1977, but improvements in the system during the intervening years, showed that the early estimate was conservative. In 2012 the number of cruisers and yachts for hire on the Canal was sixty-four, little more than one per mile, still not crowded by southern standards. It used to become crowded at Fort Augustus on Mondays but the introduction of Sunday working in the summer of 1987 solved this problem.

The provision of service facilities, such as boatyards, shops, water supplies, and refuse and sewage disposal points had to happen quickly to cope with the growth. Muirtown Basin, which has ample space for mooring boats, has become Seaport Marina overlooked by the

Jacobite Lady and *Scot II* carrying day-trippers through Dochgarroch Lock on the way to Loch Ness.

new Canal Office. Above the Locks, Caley Cruisers has a marina and chandlery beside its hire cruiser site. Fort Augustus's shops and hotels were always well placed to serve the cruiser trade because it was so easy for people to come ashore. Elsewhere new jetties now give access to historic sites, country shops and hotels. Sewage disposal was more difficult. The Highland River Purification Board, which monitors the quality of the water along the Great Glen regularly, wanted sanitary stations built urgently at Inverness, Fort Augustus and Banavie (where boats could pump their effluent ashore) and these have been provided.

In 1987 a new system of charging was introduced through the issue of licences. It recognised that fewer craft wish to pass right through the Canal by abolishing passage dues. In their place, it aimed to increase pleasure use of its waters by basing charges on the length of the boat and the intended length of stay, with charges tapering to encourage craft to extend their stay. The fee per metre for a licence for one month in 2010, for example, is a little over double the fee for eight days, which is the shortest available. The value for money also of the marina facilities for yachts on Highland canals has been emphasised, compared with mooring costs on the South Coast, as well as their extent, with the Crinan and Caledonian Canals being described as 'Britain's longest marinas'.

One unusual vessel to sail through the Canal in January 1985 was the replica of Sir Francis Drake's *Golden Hind*, but whereas the original had been the first English ship to sail right round the world, the replica ran aground several times in the Canal by not keeping to the middle. She passed through the Canal again in spring 1994 to Muirtown Basin where she was again an attraction for visitors. The *Grand Turk*, well known from the TV series *Hornblower*, also attracted crowds when she sailed through in June 2000.

Without any doubt, however, the greatest collection of sailing ships – the kind of vessels the Canal was originally built to accommodate – ever seen together on the Caledonian Canal in its long history were the thirty-seven which passed through on 28–30 July 1991. They were on their way to Aberdeen to take part in the Tall Ships Race. On the Canal, however, they were not competing but 'cruising in company' and they were enjoying each other's company. The crews, most of them young people who came from nine different countries from Ireland to the Soviet Union, shared the experience with the crowds who came to admire as their tall ships progressed gently and majestically up Neptune's Staircase in the sun. The atmosphere of carnival continued at a barbecue and ceilidh in the grounds of the Abbey at Fort Augustus and again at Muirtown Basin when the people of Inverness came to view.

Not all who enjoy the Caledonian Canal are on the water. Well over 30,000 walkers and 12,000 cyclists each year use all or sections of the Great Glen Way which extends for 79 miles between Fort William and Inverness. It came into use on 30 April 2002. They share the Canal towpaths which gives them the opportunity to experience at first hand the secret places of the Caledonian Canal, while the whole Great Glen Way opens up many of the places of interest which are described in the last section of this book.

Great Glen Week 10–17 May 2003 was a programme of outdoor events in the Glen to commemorate the bicentenary of the Caledonian Canal 1803–2003. *Sail Caledonia*, the main event aimed at sailing and

Canoes and cruisers in 2012.

rowing boats, drew twenty entries and promised to have far more in subsequent years. Sailing from west to east it provided three regattas, one at Fort William, one at Fort Augustus and one at the east end of Loch Ness, with rowing contests on the Canal reaches in between. Accompanying them on foot were thirty walkers on 'The Highland Hike' also from west to east on the Great Glen Way. They followed Canal towpaths, the old railway track and woodland paths, covering 73 miles in five days. A whole month of ceilidhs and performances by young musicians in the summer in halls and hotels, or on piers and beside the locks and even on the Canal, sailing on the *Jacobite Queen*, extended the message of the Canal bicentenary.

The Great Glen Canoe Trail has been developed at a cost of £601,000, 45% of it with EU funding. It provides 'paddler-friendly' pontoons at the locks and bridges and informal camping spots, and already in 2012 over 4,000 canoeists and kayakers are using it each year. Muirtown Basin is also a safe place for beginners. In all these ways the Canal is playing an ever-increasing part in people's enjoyment of outdoor activities in their leisure.

How have the lives of Canal workers been affected by this switch to leisure traffic? In summer locks and bridges are operated for ten hours a day, from 8 o'clock in the morning until 6:00 in the evening every day including Sundays. At this season the lives of lock-keepers from Dochgarroch in the east to Gairlochy in the west have completely changed. The idyllic quiet life a lock-keeper such as Dan Kennedy enjoyed at Cullochy has gone for ever. A modern lock-keeper who operates locks all day has little time for maintenance work, such as painting and grass-cutting. All the greater then was the achievement of Bill McLaughlan and Ian MacLaren of Gairlochy Locks who won the national award for the best-kept locks on the whole of British Waterways in 1988 and in 1997, and the Scottish award in 2003. In 2010 the Caledonian Canal won the Best Visitor Experience at the Highlands and Islands Tourism Awards and, in the same year, the Waterways Renaissance Award in the Recreation and Tourism Category.

The numbers on the Canal staff have actually increased to 39 full-time staff and 30 seasonal, due mainly to the expansion of the maintenance team. Lifelong service has become less common since the retiral of Andrew Goodwin with 49 years' service in 1986, and George Anderson with 37 years' service and Kenneth McNab with 38 (22 of them on the Caledonian) in 1987. Compared with them, only one member of the staff in 2012, Ronnie Ross with 47 years' service, and still working, was employed on the Canal in 1972 on the 150th anniversary of its opening. Old job titles, such as Keeper of the Sea-Lock and Head Lock-Keeper have been discarded – the staff are practically all Waterway operatives now. The post of Collector of Tolls, which has a historic ring going back to Andrew May in 1822, has also gone, as licence fees are simply collected at both ends of the Canal and handled by the office staff. The appointment of a woman, Mrs Jenny Rowantree, as bridge-keeper at Moy was an innovation in the early 1990s, and in 2001 Beatrice Clark was a lock-keeper on the permanent staff at Muirtown, the first woman in Scotland to hold such a position, while two women were lock-keepers for the summer season. In 2012 Toni Sutherland is on the permanent staff at Banavie Locks and other women are employed as lock-keepers in summer. In another first in November 2003 Pam Swanson became

Opposite: Canal workers at the Works Lock for the retirement presentation to length foreman George Anderson (left) which area inspector Alex Macdonald (right) made in April 1987. Sadly Alex, a former police diver and an outstanding personality whose canal responsibilities extended also to the Crinan Canal, died only four months after this at the early age of 53.

the manager of the Highland Canals, both the Caledonian and the Crinan. She was the first woman to hold these posts either together or separately, before going on to British Waterways HQ in Glasgow. She was succeeded by Russell Thomson as Waterway Manager and Andrew Ross is the Engineer based in Inverness. Overall, the changes have achieved the flexibility to respond with maximum resources at times of peak demand, and to switch staff to other work, at times of closure for example, with the compensation for permanent staff that the Canal is open for fewer hours at other times when days are shorter and customers are few.

In a major organisational change from 1 July 2012, The Caledonian Canal and other canals in Scotland ceased to be administered from day to day by the British Waterways Board and became the responsibility of Scottish Canals. Their first year's grant of £10m, came from British Waterways and a further £4.6m for named projects was given by the Scottish Government. The headquarters of Scottish Canals is in Glasgow.

It is true that in 1972 the Caledonian Canal was an amenity which was under-used. Since then walkers and cyclists have been attracted to its towpaths and canal cruising has made its waters busy with boats, and given thousands of visitors the chance to discover it. Cruising, on looking back, has probably been the ideal kind of activity to develop at this stage in the Canal's life: they could have been made for each other.

What might have been? An industrial barge-train on the Grand Canal of China . . .

. . . but a switch from industry to leisure on the Caledonian Canal as Fort Augustus welcomes the Tall Ships.

APPENDIX I
CANAL INFORMATION

Length of natural Lochs:	38 miles (61.16 km).
Length of Canal cuttings:	22 miles (35.4 km).
Total Length of Canal:	60 miles (96.56 km).
Summit level at Loch Oich:	106 feet (32.3m).
Canal is suitable for vessels up to:	150 feet long x 35 feet beam x 13½ feet draught (45.72m x 10.67m x 4.11m draught).
Corpach Basin takes vessels up to:	203 feet long x 35 feet beam x 13½ feet draught (61.87m x 10.67m x 4.11m).
Locks and bridges:	all operated by Canal staff.
No. of locks:	29.
No. of bridges:	10.
No. of pairs of gates:	42.
Time to pass a single lock:	30 minutes.
Time to pass Neptune's Staircase and Bridge:	90 minutes.
Speed limit on Canal reaches:	5 knots.
Speed on Lochs:	No limit.
Time for passage through Canal:	14 hours at least over 2 – 2½ days.
Cost of Construction (1803–22)	£912,000
Cost of Reconstruction (1843–47)	£228,000
Cost of Lock-gate mechanisation (1959–68)	£195,000
Cost of Reconstruction Fort Augustus (1995–7)	£2,894,000
Cost of Reconstruction Neptune's Staircase (1996–2001)	£5,075,000
First passage through Canal:	23–24 October 1822.

Admiralty Canal Chart:	Ref. 1791.
Licences to sail the Canal:	from sea-lock keepers at each end.
Caledonian Canal Office:	Seaport Marina Muirtown Wharf Inverness IV3 5LE Tel: 01463 233140

APPENDIX II
PLACES OF INTEREST IN THE GREAT GLEN

Inverness

A Royal Burgh created beside a royal castle in the twelfth century, Inverness is now the administrative and services capital of the Highlands with city status and a rapidly rising population. One of the bodies based here is the Highlands and Islands Enterprise which has the task, not unlike Telford's, of encouraging a variety of economic activities to provide more jobs in the Highlands. An important route-centre at the north-east end of the Great Glen, Inverness caters for large numbers of visitors who come by road, rail and air. The city has a wide selection of hotels and guest houses, as well as a large youth hostel and caravan parks, and offers a wide range of entertainment at the multi-purpose Eden Court Theatre, close to the River Ness.

Little of the ancient town remains. The Castle is recent, 1834–46, the latest in succession on the old site commanding the crossing of the River Ness. Most of the older buildings of interest are in Church Street, named from the old parish church, reconstructed in 1769–72 and attached to a solid tower dating from the fourteenth century. Flanking its gateway are two whitewashed houses which still have forestairs. Opposite is Dunbar's Hospital, 1668, once the town's grammar school, and next to it across School Lane is Bow Court, an early eighteenth-century building recently restored. Nearby is Abertarff House, an attractive restoration of an early seventeenth-century town house with a turnpike stair. Union Street and Queensgate in the town centre are well-designed corridor streets leading through to Academy Street and Station Square. Between these streets is the city's covered market.

Inverness's main attraction is the River Ness flowing through the centre of the city, crossed by four main bridges and two graceful Victorian suspension bridges for pedestrians. The red steel-span

railway viaduct replaces Joseph Mitchell's five-arch stone bridge of
1861, which collapsed in the violent floods of 7 February 1989. Views
across the River are heightened by many church steeples and towers.
Salmon fishing is popular. Upstream are Bught Park and the famous
Ness Islands, pleasant walks linked by bridges.

Within easy reach of the city to the east is *Culloden*, where the
British army under the Duke of Cumberland broke the Jacobite
forces of Prince Charles Edward Stuart in 1746 in the last land battle
fought in Britain (Museum and Information Centre). *Clava Cairns*
nearby are impressive Early Bronze Age chambered tombs, set within
circles of standing stones. Beyond Ardersier is *Fort George,* which is
well worth a visit, the finest Hanoverian fort in Britain, built after
the Battle of Culloden. To the west is *Craig Phadrig*, the vitrified fort
of the fourth century BC with magnificent views from its summit. In
the coniferous woods around it is a well-planned Forest Walk. Further
west, Telford's Lovat Bridge is named after Lord Lovat, chief of the
Frasers, whose clan area this was. The red sandstone village of *Beauly*
contains the ruins of a thirteenth-century priory. The Kessock Bridge,
replacing the old ferry, gives access to the Black Isle and the North.

Inverness to Fort Augustus – General Wade's road

The road on the south-east side of the Great Glen, A862, follows the
road built by General Wade in 1726–7 for much of the way. This
entry to the Great Glen fault is low and gentle as far as *Dores* at the
end of Loch Ness, passing through woods and arable land. South of
Dores the mountains close in on the Glen, their steep sides giving it
the appearance of a great natural canal.

Near Kinchyle, north-east of Dores are the remains of a Bronze
Age chambered cairn. Nearer to Dores is *MacBain Memorial Park*, an
enclosed garden open to visitors above the end of Loch Ness. Notable
for its plants, its rhyming requests not to take them, two splendid
wild cats in bronze on the Memorial, and impressive views over the
Loch, this is a small MacBain enclave in lands which used to belong
to Clan Chattan, a confederation of clans under the Mackintoshes,
'the clan of the cat'.

At Dores, these is a choice of roads: the high road by Loch Mhor
for distant views, and the more direct low road along the water's edge.

Inverfarigaig, where steamers used to call daily with passengers and mail, now has an interesting Forest Walk and forestry exhibition. To the north, is steep-sided *Dun Dearduil*, once a timber-laced stone-walled fort, where the stones were turned into a fused or vitrified mass by intense heat.

Foyers was also a point of call for steamers, bringing visitors to see its Falls, where the water plunges first 30 feet, then 90 feet with a deafening roar raising clouds of spray. Water diverted from above the Falls was used to produce the first hydro-electric power in Britain in 1896 for the aluminium works at Lower Foyers which were in operation until 1967. The hydro-electric scheme at Foyers today uses water from Loch Mhor to produce electricity to feed the national grid and pumps water back up from Loch Ness by its own power at off-peak periods. (There are plans for a similar type of hydro scheme at Coire Glas above Loch Lochy.) From Foyers *Meal Fuarvonie* is best seen nearly 2,300 feet high across Loch Ness.

Whitebridge, General Wade's splendid high humpbacked bridge still stands beside its successor in a pleasant spot above Fort Augustus. The road descends steeply towards Fort Augustus, giving a view that is almost aerial over the Abbey, the Rivers Tarff and Oich and the Canal.

Fort Augustus

Formerly called Kilcumein, the cell or church of St. Cumin, one of St Columba's successors as Abbot of Iona, it is roughly half-way between Fort William and the original Fort George at Inverness. *Fort Augustus* became a key point in the Government strategy of keeping the Highlands peaceful by controlling the Great Glen through forts joined by military roads, and linked with the South by new roads. The first barracks on the site of the present Lovat Arms Hotel was built soon after the 1715 Rising, and then the Fort was erected on the shores of Loch Ness between the Oich and the Tarff by General Wade. In 1746 it was captured by Jacobite forces but after the Battle of Culloden it became a base for hunting down Jacobites. It was here that the head of Roderick Mackenzie, killed in Glenmoriston because he was thought to be Prince Charlie, was brought to the Duke of Cumberland. Here too, old Simon Fraser, Lord Lovat, was a prisoner in 1746, before being taken to London for trial and execution.

The Abbey Church and Tower, Fort Augustus.

The Fort, reconstructed after 1746, was capable of holding a garrison of three hundred men but in later years when the threat of insurrection was over, many of the men were military veterans. In 1773 Dr Johnson and James Boswell were well received here, 'in good company and with a good supper before us'. After the Napoleonic Wars the soldiers left, their equipment being one of the first cargoes carried from Fort Augustus through the Canal. Bought by Lord Lovat in 1867, the land and the Fort were donated to the Benedictine Order of monks who built the Abbey and School by 1880, but the School closed in 1993 and the Abbey itself four years later. Since then it has changed hands more than once and most of it turned into flats, described as luxury apartments. Almost nothing of the Fort remains but the village retains the name.

In contrast to Gordon Cumming's collection of big game trophies, the tourist attraction of Fort Augustus a century ago, the Caledonian Canal Heritage Centre, which faces the Canal locks, is well worth a visit. Visitors interested in railway history can trace the remains of the old line from Spean Bridge and see the pillars of the former railway bridge across the Oich.

Inverness to Fort Augustus (NW side of Loch Ness)

The main road, A82, constructed in the 1930s, passes along the west side of the River Ness and over the Canal at Tomnahurich Bridge to follow the line of the road built by Donald McKintosh for Thomas Telford. By Dochfour Loch, the road is the reinforced towpath built in the 1840s. At Dochgarroch where there is a caravan park, the lock is easily reached from the main road. At Lochend, Loch Ness is reached, and at several places along the road there are good viewing points along the Loch. On the right a steep road leads up to *Abriachan*, once a thriving crofting settlement, beautifully situated above Loch Ness.

The village of *Drumnadrochit* nestles on low land at the mouth of Glen Urquhart, overlooking Urquhart Bay. It is a popular holiday centre for fishermen, walkers, pony-trekkers and everyone curious to find out more about the Loch Ness Monster. The road up Glenurquhart passes the Clava-type chambered cairn at *Corrimony* on the way to Cannich near the head of Strath Glass, once the clan lands of the Chisholms. Above Cannich are *Fasnakyle* power station and lonely *Glen Affric*. Glenurquhart Highland Games attracts large crowds to Drumnadrochit at the end of August.

In a commanding position on a promontory overlooking Loch Ness stand the ruins of *Urquhart Castle* on the site of an early vitrified fort. In the Wars of Independence the Castle was important for control of the Great Glen. It was in the hands of Edward I of England for a time and was later taken by Robert the Bruce. In the reign of James IV it passed to the Grant family and today it is under the guardianship of Historic Scotland and attracts 250,000 visitors a year.

To the south is the cairn which commemorates John Cobb's death during his attempt on the world waterspeed record here in 1952. At *Invermoriston* the road to Kyle of Lochalsh and Skye up Glen Moriston passes Roderick MacKenzie's Cairn and Loch Cluanie into *Glen Shiel* where the only fighting in the 1719 Rising took place. On the south side of Loch Duich an exciting road over *Mam Rattachan* (over 1,100 ft) leads to *Glenelg* where there are two well-preserved brochs, circular defensive towers of the Iron Age found only in Scotland.

The old bridge and the falls at Invermoriston are worth visiting in a wooded glen. All along the Great Glen the lower slopes of the hills are clothed with coniferous trees. *Inchnacardoch* forest near Fort Augustus, with its own little village at Jenkins Park, was the first state forest in Scotland. The Forestry Commission's policy of providing walks for visitors through the woods with information at points of interest has resulted in two Forest Walks here, one along the banks of the Oich, the other up *Tor Dhuin* with outstanding viewpoints towards Loch Oich, over Kytra lock and north-east to Fort Augustus. Tor Dhuin is a vitrified fort like Craig Phadrig.

Fort Augustus to Spean Bridge

South of Fort Augustus, General Wade's military road to Badenoch, constructed in 1731, climbs up above the River Tarff through the *Corrieyairack* pass at two and a half thousand feet. The road linked this Fort in the middle of the Great Glen with the Ruthven Barracks near Kingussie and the road south at Dalwhinnie. Although it was intended for use by Government troops, Prince Charles Edward's Highland army gained the greatest military advantage from it when they advanced along it at the beginning of their campaign in 1745. They had been joined by the MacDonells of Glengarry at the head of Loch Oich by Aberchalder House, where the main road now crosses the Canal.

Invergarry Castle, the domain of the chiefs of the MacDonells of Glengarry, is in ruins, having suffered destruction by Government forces twice in the seventeenth century and again after the Battle of Culloden. It was Alastair Ronaldson MacDonell of Glengarry who erected the Monument at the *Well of the Heads* to commemorate gory redress for the murder of the young chief of the MacDonells of Keppoch. The road west from Invergarry joins the Glenmoriston road to Kyle of Lochalsh.

Returning to the south-east side of the Glen at Laggan swing-bridge, the entrance to the Great Glen Water Park is on the left. Then the road passes below South Laggan forest and runs close to the shore of Loch Lochy until it sweeps inland at Invergloy to join the Wade road near *Low Bridge* where the military road crossed the River Gloy. Ahead is the impressive *Commando Memorial* to their members who died in the Second World War (see p. 15). South-west of it are the ruins of General Wade's *High Bridge* over the Spean, where a small band of Highlanders defeated soldiers of the Royal Scots in the first fighting of the 'Forty-five.

The bridge at *Spean Bridge* was built by Telford in 1819. At *Roy Bridge*, where another Telford bridge has been made redundant by modern road improvements, a rough winding side-road leads up to the *Parallel Roads of Glen Roy*, where an indicator-board and car park have been provided. For a long time, the Parallel Roads were thought to have been man-made. As late as 1813 J. Robertson in *General View of the Agriculture of the County of Inverness* wrote:

> The Parallel Roads are a great curiosity and must have been a work of immense labour. The maker was unquestionably a personage of great influence and authority who could command the labour of such a multitude of people as this undertaking seems to have required . . . The only useful purpose for which these roads seem to have been made was hunting: . . . the roads of Glen Roy facilitated the access of archers to the deer . . . and also rendered it convenient for the spectators, whether on foot or on horseback, whether ladies or gentlemen, to follow the sport wherever it was.

The 'roads' are in fact natural, not only parallel to each other, but level at 1,149 ft, 1,068 ft and 857 ft. They were the beaches of lakes formed when the exit from Glen Roy was blocked by a glacier towards the end of the last Ice Age. The retreat of the ice barrier explains the changes in level and at the lowest level the lake extended into Glen Spean. Glen Gloy also has parallel roads.

Spean Bridge to Corpach

The minor road west below the Commando Memorial leads over *Mucomer Bridge* to Gairlochy. The Bridge over the new cutting for the Lochy was made by Telford's men. A modern power station has been built and harnesses the Falls. At Mucomer Farm lived A.A. Cameron, one of the giants of heavy events at Highland Games before the First World War. His supremacy was acknowledged by one writer who referred to 'the exponents of hammer, ball and caber from A.A. Cameron downwards . . .' Nearby, John Graham of Claverhouse, 'Bonnie Dundee', raised an army of Highlanders for James II in a campaign which ended with his death in spite of victory at Killiecrankie in 1689.

Lochaber is Cameron country, where the clan had gradually extended its authority and had often to fight to retain its lands. The Camerons had a reputation for cattle-thieving, a dangerous trade but for men living in such infertile country a necessary part of their way of life. Place-names like Achnacarry, Clunes, Erracht, Fassifern which ring in clan history, persist in a region little changed by the passage of time. The 'Gentle Lochiel's' *Achnacarry Castle* was destroyed by Government soldiers because of his part in the 'Forty-five. Its successor, the home of Cameron of Lochiel the clan chief, was the Commando training centre in the Second World War. West of Clunes, the road passes through the *Dark Mile*, once a track through dense woods. Above it Prince Charles Edward was sheltered by Cameron of Clunes when troops were searching for him after Culloden. The area is full of associations with his wanderings during the five months he was on the run. Eventually he passed along the shores of Loch Arkaig and over to Loch nan Uamh to embark on a ship for France. *Glenfinnan* (Monument and Information Centre) where his standard was raised at the beginning of the Rising is not far from the head of Loch Eil. Less than fifty years later, Alan Cameron of Erracht raised the Cameron Highlanders to fight *for* the Government in 1793.

The road north from Gairlochy was undertaken by Colonel Cameron of Locheil in 1809 and the bridges were built by Simpson and Wilson's masons. The road south to Corpach, the work of the Canal contractors, is the best for exploring the Canal and passing under the aqueducts over the *River Loy* and the *Shangan Burn*. On the other side of the Loy aqueduct on the right was a sawmill driven by a dam from the Loy when the Canal was under construction. The Shangan Burn aqueduct gives access to the ruins of old *Tor Castle* by the River Lochy. The castle has been interpreted variously as an ancient capital of the Scots and the castle of Banquo, Macbeth's ally and later victim in Shakespeare's play. There is an avenue of trees near the River called Banquo's Walk. Tor Castle was certainly the stronghold of the Cameron chiefs in the sixteenth and seventeenth centuries.

At Banavie are the eight locks in *Neptune's Staircase*. On Corpach Moss is the post-war village of *Caol* which grew so rapidly that its population (over 4,000) is as great as neighbouring Fort William. To the west *Corpach* at the mouth of the Canal also expanded as a result of employment created by the erection of the Pulp and Paper Mill at *Annat Point* in 1966. Attracted to this site by a combination of advantages – ample supplies of soft water, the prospect of increasing supplies of softwood from Highland forests, and the existence of road and rail communications, the Mill was one of the pioneers of

Tall ships ascending Neptune's Staircase in July 1991, original lock-keepers' houses on the left.

large-scale industry in the Highlands but was reduced in scale to paper-making only in 1980.

The road along the north side of Loch Eil leads past Glenfinnan to Arisaig and Mallaig and the sands of Morar.

Spean Bridge to Fort William

The main road passes what was once the Great Glen Cattle Ranch, an experiment in land reclamation and large-scale cattle-rearing undertaken by Mr J.W. Hobbs between 1945 and 1961. The modern Inverlochy Castle is a Victorian building where Queen Victoria spent a week in 1873 before sailing along the Caledonian Canal on her return to Balmoral. The Castle is now a hotel with a high reputation for its meals.

Enough of the original *Inverlochy Castle* remains to show that it was a place of strength at the western entrance to the Great Glen. Built in the late thirteenth century, it is square, defined by curtain-walls 10 feet in thickness, with round towers at each corner, and surrounded by a moat fed by the River Lochy. Near it in 1645 Montrose's Highlanders

and 'Wild Irishmen' overwhelmed Argyll's Campbells in a bid to restore Charles I's fortunes in Scotland at the time of the English Civil War.

Ben Nevis looms over Lochaber. It is 4,406 ft high, the highest mountain in Britain. Walkers may follow a path to the summit, taking up to eight hours for the return trip. Climbers are attracted to the cliff face on the north-east side. Those who venture on the Ben are warned of the dangers of cold and mist. A race up and down Ben Nevis is held annually in September. On the slopes of *Aonach Mor* (3,999 ft) east of Ben Nevis, a well-equipped and popular skiing centre has been established.

Fort William

Now a popular tourist centre for holidays in Lochaber, *Fort William* was founded as a fort called Inverlochy by General Monck in 1654. After the accession of William and Mary, the fort was rebuilt in 1690. It was called Fort William, and the village outside it was named Maryburgh after the only joint monarchs in British history. The Fort had links with the Massacre of Glencoe and the Appin Murder. It withstood a Jacobite siege in 1746 but succumbed to the West Highland Railway in 1890, when engine sheds took the place of barrack blocks, but a gateway of the old Fort has been rebuilt at the entrance to the Craigs burial-ground (see page 13). Although the Fort has gone, the name survives as the name of the town.

Strung along the side of Loch Linnhe, Fort William is well supplied with hotels, restaurants and shops. The West Highland Museum is of particular interest to visitors and the town also has a whisky distillery. To relieve traffic congestion in a town built on a narrow restricted site, the River Nevis has been diverted to join the Lochy and its channel filled in to allow the land called An Aird to be used for a relief road and a new site for the rail station.

Up Glen Nevis is the hill-fort called *Dun Dearduil*, also vitrified like the fort with the same name at Inverfarigaig. Above it among some of the finest wild scenery in Britain are the Falls of Nevis and the 400 ft *Nevis Gorge*.

Opposite: On the 300th anniversary of the foundation of Fort William, HM Queen Elizabeth talking to lock-keepers at Corpach sea-lock.

The main road south from Fort William reaches *Ballachulish* in a district once famous for its slate quarries. From South Ballachulish there is a choice of routes, along the shores of Loch Linnhe to Oban or through the mountains by Glencoe where memories still linger of the Massacre in 1692.

FURTHER READING

Manuscripts

Ministry of Transport Records on Caledonian and Crinan Canals:
  Letters, Minute Books, Proceedings of the Commissioners
  (National Archives of Scotland)
Papers on Highland Roads and Bridges, Caledonian Canal etc.,
  1803–51 (House of Lords Record Office)
Notebooks of Thomas Telford (Institution of Civil Engineers)
Caledonian Canal MSS. (Institution of Civil Engineers)
Journal of George Brown (National Library of Scotland)
Letter Book of John Telford (Caledonian Canal Office)

Plans

Atlas to the Life of Thomas Telford, Civil Engineer
Original Drawings by Thomas Telford (Institution of Civil
  Engineers).
Plans by George Brown and W. Cuming (National Archives of
  Scotland).
Morrison's Survey of the Caledonian Canal, 1843–4 (National
  Archives of Scotland).
Thomas Rhodes's Drawing Book (Caledonian Canal Office).
Plans by George Brown, W. Cuming, William Jessop and others
  (Highland Council Archive)

Reports

Caledonian Canal Commissioners, 1803–1920.
Conditions in the Highlands and Islands and the Practicability of
  Relief by Emigration, 1841.

Royal Commission on Canals and Waterways, 1906–8.
Canals and Waterways, 1955.
Inland Waterways, 1958.
Fraenkel Report, 1975.

Articles

M.I. Adam, The Causes of the Highland Emigrations of 1783–1803
  (*Scottish Historical Review*, vol. xvii, 1920).
The Caledonian Canal (*Blackwood's Magazine*, 1820).
Samuel Hughes, Memoir of William Jessop (*Weale's Engineering
  Papers*, 1844).
Alexander Ross, The Caledonian Canal and its Effects on the
  Highlands (*Transactions of the Gaelic Society of Inverness*, vol. xiii,
  1886–7).

Books on Telford and Canals

Anthony Burton, *Thomas Telford*, 1999.
A.D. Cameron, *Thomas Telford and the Transport Revolution*, 1979.
Sir Alexander Gibb, *The Story of Telford*, 1935.
Charles Hadfield, *The Canal Age*, 1968.
Charles Hadfield, *Thomas Telford's Temptation*, 1993
Charles Hadfield and A.W. Skempton, *William Jessop, Engineer*,
  1979.
A.R.B. Haldane, *New Ways Through the Glens*, 1962.
Guthrie Hutton, *Caledonian – The Monster Canal*, 1992.
Jean Lindsay, *The Canals of Scotland*, 1968.
A. Penfold (ed.), *Thomas Telford: Engineer*, 1980.
E.A. Pratt, *Scottish Canals and Waterways*, 1922.
L.C.T. Rolt, *Thomas Telford: A Biography*, 1958.
Thomas Telford, *Life of Thomas Telford, Civil Engineer, written by
  himself*, ed. J. Rickman, 1838.

Other Useful Books

James Barron, *The Northern Highlands in the Nineteenth Century*,
  3 vols., 1903–13.

Duckworth and Langmuir, *West Highland Steamers*, 1967.

David Duff, *Victoria in the Highlands*, 1968.

Donald B. MacCulloch, *Romantic Lochaber, Arisaig and Morar*, 3rd edn, 1971.

James Miller, *The Dam Builders: Power from the Glens*, 2002.

Joseph Mitchell, *Reminiscences of My Life in the Highlands*, 2 vols., 1883.

Robert Southey, *Journal of a Tour in Scotland*, 1819.

John Thomas, *The West Highland Railway*, 1965.

Charles Richard Weld, *Two Months in the Highlands, Orcadia and Skye*, 1860.

Orlo Williams, *Life and Letters of John Rickman*, 1912.

INDEX